T0353469

RELAXATION TECHNIQUES FOR THE SIMULATION OF VLSI CIRCUITS

THE KLUWER INTERNATIONAL SERIES
IN ENGINEERING AND COMPUTER SCIENCE

VLSI, COMPUTER ARCHITECTURE AND
DIGITAL SIGNAL PROCESSING

Consulting Editor

Jonathan Allen

Other books in the series:

Logic Minimization Algorithms for VLSI Synthesis, R.K. Brayton, G.D. Hachtel, C.T. McMullen, and A.L. Sangiovanni-Vincentelli. ISBN 0-89838-164-9.

Adaptive Filters: Structures, Algorithms, and Applications, M.L. Honig and D.G. Messerschmitt. ISBN 0-89838-163-0.

Computer-Aided Design and VLSI Device Development, K.M. Cham, S.-Y. Oh, D. Chin and J.L. Moll. ISBN 0-89838-204-1.

Introduction to VLSI Silicon Devices: Physics, Technology and Characterization, B. El-Kareh and R.J. Bombard. ISBN 0-89838-210-6.

Latchup in CMOS Technology: The Problem and Its Cure, R.R. Troutman. ISBN 0-89838-215-7.

Digital CMOS Circuit Design, M. Annaratone. ISBN 0-89838-224-6.

The Bounding Approach to VLSI Circuit Simulation, C.A. Zukowski. ISBN 0-89838-176-2.

Multi-Level Simulation for VLSI Design, D.D. Hill, D.R. Coelho. ISBN 0-89838-184-3.

RELAXATION TECHNIQUES FOR THE SIMULATION OF VLSI CIRCUITS

by

Jacob K. White
Thomas J. Watson Research Center
IBM Corporation

and

Alberto Sangiovanni-Vincentelli
University of California
Berkeley, California

KLUWER ACADEMIC PUBLISHERS
Boston/Dordrecht/Lancaster

Distributors for North America:
Kluwer Academic Publishers
101 Philip Drive
Assinippi Park
Norwell, MA 02061, USA

Distributors for the UK and Ireland:
Kluwer Academic Publishers
MTP Press Limited
Falcon House, Queen Square
Lancaster LA1 1RN, UNITED KINGDOM

Distributors for all other countries:
Kluwer Academic Publishers Group
Distribution Centre
Post Office Box 322
3300 AH Dordrecht, THE NETHERLANDS

Library of Congress Cataloging-in-Publication Data

White, Jacob K.
  Relaxation techniques for the simulation of VLSI circuits.

  (The Kluwer international series in engineering and computer science ;
SECS 20. VLSI, computer architecture, and digital signal procession)
    Bibliography: p.
    Includes index.
    1. Integrated circuits—Very large scale integration—Mathematical models.
  2. Integrated circuits—Very large scale integration—Data processing.
  3. Relaxation methods (Mathematics) I. Sangiovanni-Vincentelli, Alberto. II. Title.
  III. Series: Kluwer international series in engineering and computer science ; SECS 20)
  IV. Series: Kluwer international series in engineering and computer science.
  VLSI, computer architecute, and digital signal processing.
  TK7874.W48 1986   621.395   86-20950
  ISBN 0-89838-186-X

Copyright © 1987 by Kluwer Academic Publishers

All rights reserved. No part of this publication may be reproduced, stored in a retrieval system, or
transmitted in any form or by any means, mechanical, photocopying, recording, or otherwise,
without the prior written permission of the publisher, Kluwer Academic Publishers, 101 Philip
Drive, Assinippi Park, Norwell, MA 02061.

Printed in the United States of America

CONTENTS

PREFACE

Circuit simulation has been a topic of great interest to the integrated circuit design community for many years. It is a difficult, and interesting, problem because circuit simulators are very heavily used, consuming thousands of computer hours every year, and therefore the algorithms must be very efficient. In addition, circuit simulators are heavily relied upon, with millions of dollars being gambled on their accuracy, and therefore the algorithms must be very robust.

At the University of California, Berkeley, a great deal of research has been devoted to the study of both the numerical properties and the efficient implementation of circuit simulation algorithms. Research efforts have led to several programs, starting with CANCER in the 1960's and the enormously successful SPICE program in the early 1970's, to MOTIS-C, SPLICE, and RELAX in the late 1970's, and finally to SPLICE2 and RELAX2 in the 1980's.

Our primary goal in writing this book was to present some of the results of our current research on the application of relaxation algorithms to circuit simulation. As we began, we realized that a large body of mathematical and experimental results had been amassed over the past twenty years by graduate students, professors, and industry researchers working on circuit simulation. It became a secondary goal to try to find an organization of this mass of material that was mathematically rigorous, had practical relevance, and still retained the natural intuitive simplicity of the circuit simulation subject. To meet these two goals, we decided to present a description of the entire circuit simulation process, from formulation of the circuit differential equations to numerical methods for solving those equations, with a strong emphasis on those issues relevant to relaxation algorithms. We have, however, presumed some knowledge of linear algebra and functional analysis. A more-than-adequate introduction can be found in *Real Analysis* by W. Rudin or *Iterative Solutions of Nonlinear Equations in Several Variables* by J. M. Ortega and W. C. Rheinbolt.

Another goal of this book is to act as a guide to the computational and implementation issues in the development of relaxation-based circuit simulators. In particular, the program RELAX2.3 is used as an example of a waveform relaxation-based circuit simulator. RELAX2.3 is available from the Industrial Liaison Program Software Distribution Office of the Department of Electrical Engineering and Computer Sciences of the University of California, Berkeley.

We would like to acknowledge the many fruitful discussions with Richard Newton and Albert Ruehli on circuit simulation. We would like to thank Barbara Bratzel, Giovanni DeMicheli, Ken Kundert and Don Webber for proof-reading the manuscript, Resve Saleh for providing several of the examples, and Farouk Odeh for assisting in the proofs of most of the theorems. We would also like to thank Carl Harris for his constant encouragement and support. The research that has led to the results presented here has been supported by a number of government agencies and companies: DARPA under Contract N00039-83-C-0107, the Joint Electronic Service under contract F49620-84-C-0057, the MICRO program of the State of California, IBM, Harris Semiconductors, Phillips, Intel, SGS Microelectronica, GTE, and Hughes Aircraft.

Finally, we would like to acknowledge the love and patience of our families, to whom this book is dedicated. Without their understanding and moral support, we would not have been able to devote as much of our energy to research and writing.

RELAXATION TECHNIQUES FOR THE SIMULATION OF VLSI CIRCUITS

CHAPTER 1 - INTRODUCTION

SECTION 1.1 - SIMULATION FOR IC DESIGN

Simulation programs have completely replaced *breadboards* (prototype boards with discrete components) as a technique for verifying integrated circuit design acceptability. In fact, a breadboard may give results which have little resemblance to the manufactured circuit performances due to their completely different parasitic components. Simulation programs are also used to avoid the fabrication of prototype integrated circuits because fabrication and testing of an integrated circuit to verify a design is both expensive and time-consuming. Moreover, extensive probing is not possible and modification of circuit components to determine a better design is practically infeasible.

Today, many different forms of simulation can be used for the verification of large digital integrated circuit designs at the various stages of the design process. They may be classified as *behavioral* (also called algorithmic or functional) simulators, *register-transfer-level (RTL)* simulators, *gate-level-logic* simulators, *timing* simulators, *circuit* simulators, *device* simulators, and *process* simulators.

Behavioral simulators are used at the initial design phase to verify the *algorithms* of the digital system to be implemented. Not even a general structure of the design implementation is necessary at this stage. Behavioral simulation might be used to verify the communication protocols in a multiprocessor design, for example.

Once the algorithms have been verified, a potential implementation *structure* is chosen. For example, a microprocessor, some memory, and some special-purpose logic may be chosen to implement the communication protocol mentioned above. An RTL simulator can be used to verify the design at this level. Since the exact circuit parasitics and other implementation details are not yet known, RTL simulators provide only crude timing information, but can provide the designer enough information to make decisions about hardware/firmware trade-offs. A variety of RTL languages and associated simulators have been described in the literature [71,72,73].

Depending on the design methodology and certain technology issues, a gate-level design may be undertaken, in which each of the RTL modules is further partitioned into low-level logic building blocks, or gates. A logic simulator may then be used to verify the design at this level. Sophisticated delay models may be introduced and testability analyses performed.

From the gate level, transistors and associated interconnections are generated to implement the design as an integrated circuit. Accurate electrical analysis can be performed for small parts of this design using a circuit-analysis program [2,3] or larger blocks may be analyzed in less detail using a timing simulator[7,74]. Once the integrated circuit layout is complete, accurate circuit parameters, such as parasitic-capacitance values and transistor sizes, may be extracted. The extracted data may then be used to improve the accuracy of the predicted performance. Extracted data can be used to improve delay models at both the gate and RTL levels, or used directly in an electrical analysis program.

Finally, generating transistor models for circuit analysis, and designing or tuning a fabrication process is aided by device and process simulators. Process simulators use as inputs the control parameters of the process, such as furnace temperature and initial impurity densities, and produce physical information such as doping profiles and oxide thicknesses. These outputs are then used as inputs to device simulators, which are used to determine device characteristics such as transistor currents or diffusion capacitances.

A number of simulators have been developed recently which span a range of these levels in the simulation hierarchy. These simulators are called *mixed-level* simulators [75,76,77,78] and allow different parts of a circuit to be described at different levels of abstraction. This approach permits a smooth transition between different levels of simulation.

SECTION 1.2 - CIRCUIT SIMULATION

The evolution of IC processing technology and the quest for improved performance of IC designs are changing the relative use of the tools included in the hierarchy described above. With the present technology, entire complex systems can be integrated on a single chip and, while higher-level tools such as gate-level-logic simulators and switch-level simulators can verify circuit function and provide first-order timing information, only the accurate analysis provided by circuit simulation can verify the performance of the design. Crucial parts of the design simply have to be simulated at the detailed circuit level. In addition, memory design has become so aggressive that circuit simulation is essential to verify the functionality of the part.

Two commonly used circuit simulators, SPICE[2] and ASTAP[3], perform a variety of analyses, including dc, ac, and time-domain transient analysis of circuits containing a wide range of nonlinear, active circuit devices such as field-effect and bipolar-junction transistors. The most common analysis performed by circuit simulators, and the most expensive in terms of computer time, is time-domain transient analysis. This analysis is performed to obtain electrical waveform information which can be extremely accurate if the device models and parasitics of the circuit are well characterized.

Conventional circuit simulators like SPICE and ASTAP were designed in the early 1970's when the goal was the cost-effective analysis of circuits containing fewer than a hundred transistors. Because of today's need to verify the performance of larger circuits, many designers have used these programs to simulate circuits containing thousands of transistors even though the computation requires several CPU hours. In some companies, the simulation of circuits containing many thousands of devices is performed routinely and at great expense. For example, at one major IC house SPICE is run more than 10,000 times per month. At another major IC house, more than 70% of an IBM 3090 is devoted to circuit simulation. Circuits containing as many as 10,000 active devices have been simulated with circuit simulators. For some of these circuits, running times on the order of one hour of IBM 3081 CPU *per timepoint* (!) have been reported.

SECTION 1.3 – STANDARD CIRCUIT SIMULATORS

Standard circuit simulators, such as SPICE and ASTAP, use standard, or direct, techniques based on the following four steps:

i) An extended form of the nodal-analysis technique to construct a system of the differential equations from the circuit topology.

ii) Stiffly stable implicit integration methods, such as the backward-difference formulas, to convert the differential equations which describe the system into a sequence of nonlinear algebraic equations.

iii) Modified Newton methods to solve the algebraic equations by solving a sequence of linear problems.

iv) Sparse Gaussian elimination to solve the systems of linear equations generated by the Newton method.

There are two reasons why the direct approach described above can become inefficient for large circuits. The most obvious reason is that sparse-matrix solution time grows super-linearly with the size of the problem. Experimental evidence indicates that the point at which the matrix solution time begins to dominate the computation involved in circuit simulation is when the system has more than several thousand nodes, and this is the size of systems that are beginning to be simulated for new IC designs.

The direct methods become inefficient for large circuits also because, for large differential equation systems, the different circuit variables are changing at very different rates. Direct application of the integration method forces every differential equation in the system to be discretized identically, and this discretization must be fine enough so that the fastest-changing circuit variable in the system is accurately represented. If it were possible to pick different discretization points, or timesteps, for each differential equation in the system so that each could use the largest timestep that would accurately reflect the behavior of its associated circuit variable, then the efficiency of the simulation would be greatly improved. This is particularly important in the case of large digital circuits, where most of the circuit variables are inactive or *latent* that is, for most of the time-points most of the circuit variables are not changing at all.

SECTION 1.4 - RELAXATION-BASED CIRCUIT SIMULATORS

Several modifications of the direct method have been used that both avoid solving large sparse matrices and allow the individual equations of the system to use different timesteps [4,5,6,7,8,9,10,11]. In addition, most of them exploit another key characteristic of large-scale circuits designed in the metal-oxide-semiconductor(MOS) technology: *unidirectionality.* MOS transistors are essentially unidirectional because the gate of the device is insulated with respect to the drain and the source of the device. This physical property implies that current through the gate is independent of the voltages at the other device terminals, if the effects of small gate-to-drain and gate-to-source capacitances are ignored.

In this book, we focus our attention on relaxation-based techniques for the transient simulation of large-scale MOS integrated circuits. Although relaxation techniques have been known for centuries, these techniques were first used in a circuit simulator in the program MOTIS[7] in 1975. MOTIS used the Jacobi-semi-implicit integration method (See Section 3.4). This technique, coupled with table-look-up transistor models, is up to two orders of magnitude faster than standard circuit simulation techniques, and is accurate enough to verify the timing of signals. For this reason, MOTIS was referred to as a timing simulator.

In order to improve the accuracy of the results computed using the MOTIS program, modified algorithms were developed and used in the MOTIS-C program[79]. In particular, MOTIS-C used a variant of the Seidel-semi-implicit integration algorithm (See Section 3.4). The MOTIS-C program was as fast as the MOTIS program, and sometimes, but not always, more accurate. In particular, using the MOTIS-C program, it was observed that the accuracy of the Seidel-semi-implicit integration algorithm is a function of the order in which the equations are processed.

The Seidel-semi-implicit integration algorithm exploits the almost unidirectionality of digital MOS transistor circuits, and can be more accurate than the Jacobi-semi-implicit integration algorithm for those circuits, if the flow of the signals through the circuit is followed correctly. This was exploited in the program SPLICE[76] which used the Seidel-semi-implicit algorithm, and introduced an event-driven approach to ordering the equations. In MOTIS-C the equations for the Seidel-semi-implicit method were ordered arbitrarily (in fact, the equations were processed in the order the user entered the nodes), and in SPLICE the equations were ordered dynamically, by trying to follow the flow of the signal in the circuit. In addition, SPLICE introduced a selective-trace algorithm which exploited circuit latency by automatically skipping inactive circuit nodes during the integration process.

For circuits where feedback paths existed and nodes were tightly electrically coupled, the Gauss-Jacobi and Gauss-Seidel semi-implicit integration algorithms were frequently inaccurate and introduced oscillations into the solutions. A theoretical analysis of these semi-implicit techniques showed that they had limitations due to numerical instability[6]. Various methods of improving the stability of the Gauss-Seidel and Gauss-Jacobi semi-implicit integration algorithms were investigated; in particular techniques based on symmetric displacement were examined[38,6].

Although semi-implicit methods with better stability properties than the Gauss-Seidel and Gauss-Jacobi semi-implicit methods were uncovered, they still had limitations due to instability. It was also shown that Gauss-Seidel-Newton algorithms(see Section 3.3) had guaranteed convergence properties when applied to MOS circuits given very mild assumptions. For this reason, SPLICE was altered to use the implicit-Euler integration method combined with a modified Gauss-Seidel-Newton algorithm(see Section 3.3). The event-driven selective-trace version of the Gauss-Seidel-Newton algorithm was called iterated timing analysis (ITA).

While the ITA algorithm was being developed, it was also noted that relaxation techniques were available for the solution of linear and nonlinear algebraic equations but that no relaxation technique had been tried *directly* on the system of differential equations describing the circuit. Waveform relaxation (WR) (See Chapter 4)[11,12,13,14,15,16,17,18] was then proposed, based on "lifting" the Gauss-Seidel and Gauss-Jacobi relaxation techniques for solving large algebraic systems to the problem of solving the large systems of ordinary differential equations associated with MOS digital circuits. It was also shown that WR had guaranteed convergence properties when applied to MOS circuits given very mild assumptions. An experimental program for the simulation of MOS digital circuits, RELAX, was developed to test the concept of WR, and produced promising results.

Theoretical and practical results on WR[17] showed that for some circuits, convergence can be quite slow. Techniques to avoid these problems were developed[18] and a new version of RELAX, called RELAX2, was written. Several other WR-based circuit simulators were developed, for example TOGGLE[15] at IBM and SWAN at ESAT at Leuven.

On a uniprocessor, relaxation-based programs can show speed improvements over direct methods of up to an order of magnitude for large problems. In addition, both ITA and WR are particularly amenable to use on multiprocessors because the computational methods already decompose the problem. A distributed form of the ITA algorithm, called DITA, has been recently developed and a prototype DITA simulator, the MSPLICE program, has been implemented[34]. A parallel version of the waveform relaxation simulator, RELAX2, has also be implemented and tested[31,66].

SECTION 1.5 - NOTATION

We will use the following standard notation throughout the text:

$\mathbb{R}$ $\equiv$ The set of real numbers.

$\mathbb{C}$ $\equiv$ The set of complex numbers.

$\mathbb{R}^n$ $\equiv$ The set of n − length real vectors.

$\mathbb{C}^n$ $\equiv$ The set of n − length complex vectors.

$\mathbb{R}^{n \times m}$ $\equiv$ The set of real matrices with n rows and m columns.

$x \in \mathbb{R}^n$ $\equiv$ The variable x is an n −length real vector.

x_i $\equiv$ The ith element of a vector x.

$A \in \mathbb{R}^{n \times m}$ $\equiv$ The matrix A a real matrix with n rows and m columns.

a_{ij} $\equiv$ The ij^{th} entry of a matrix A.

A^T $\equiv$ The transpose of the matrix A , that is $(a_{ij})^T = a_{ji}$.

$\rho(A)$ $\equiv$ The spectral radius of the matrix A , that is, the maximum value of the magnitudes of the eigenvalues of A.

$|a| \equiv$ The magnitude of the scalar a.

$\|x\| \equiv$ Any norm of the vector x

$f:\mathbb{R} \to \mathbb{R} \equiv$ The function f maps real numbers into real numbers.

$\max_x f(x) \equiv$ The maximum value achieved by $f(x)$ over the range of x.

$\dfrac{\partial f}{\partial x}(\tilde{x}) \equiv$ The derivative of the function f with respect x, evaluated at the point $\tilde{x}$.

$\|x\|_\infty \equiv$ The norm of the vector x given by $\max_i |x_i|$

$\|f\| \equiv \max_{x \neq 0} \dfrac{\|f(x)\|}{\|x\|}$

$[0,T] \equiv$ The set of real numbers greater than or equal to zero and less than or equal to T

$C([0,T], \mathbb{R}^n) \equiv$ The set of all continuous $f[0,T] \to \mathbb{R}^n$

$\displaystyle\sum_{i=1}^{n} x_i \equiv$ The sum $x_1 + x_2 + ... + x_n$.

$\displaystyle\prod_{i=1}^{n} x_i \equiv$ The product $x_1 x_2 ... x_n$.

CHAPTER 2 - THE CIRCUIT SIMULATION PROBLEM

As mentioned in the introduction, circuit simulation amounts to solving numerically the system of nonlinear ODE's that describes the dynamic behavior of a circuit. In this chapter, we will address three topics: the construction of a system of differential equations from the description of a circuit, the mathematical properties of the constructed systems, and the issues to consider when choosing a numerical method for solving the system.

SECTION 2.1 - FORMULATION OF THE EQUATIONS

The behavior of an electrical circuit can be accurately described by a set of equations involving node voltages, currents, charges and fluxes. The basic equations are formed from the Kirchoff Voltage Law (KVL), the Kirchoff Current Law (KCL), and the constitutive or branch equations[36]. The constitutive or branch equations are mathematical representations of the relations between voltages, currents, charges and fluxes at the terminals of the devices forming the circuit. These equations depend on the physics of the devices and are often highly nonlinear. The KVL and KCL equations are independent of the detailed equations that describe the individual devices in a circuit, but are determined entirely from the way in which the devices are connected together, which we refer to as the circuit topology. For our purposes, we will consider the Kirchoff Current Law as stating that for any circuit node, the algebraic sum of currents incident at a node must equal the rate of change of the algebraic sum of charges at a node, and the Kirchoff Voltage Law as stating that the sum of the voltages across the devices in any closed loop must be zero.

SECTION 2.1.1 - Branch Equations

The branch equations can be grouped into three basic sets: resistive, capacitive and inductive. Resistive equations relate voltage to current, capacitive equations relate voltage to charge, and inductive equations relate current to flux.

Branch equations for an n -terminal resistive device can be represented by a set of $(n - 1)$ algebraic equations involving $(n - 1)$ terminal voltages and currents. One terminal is used as a reference node, and the voltages of the other $(n - 1)$ terminals are determined with respect to this reference node. From KCL and the fact that there are no charges, it follows that the sum of the n device terminal currents is zero, and therefore any one device terminal current is a linear combination of the other $(n - 1)$ currents. In geometrical terms, the equations can be represented by a surface in a $2(n - 1)$-dimensional space where the axes represent all but one of the device currents and all of the terminal voltages except the reference. For example, consider the diode in Fig. 2.1. The current through the diode i_a can be computed (to first order) from the following equation:

$$i_a = I_s(e^{v_{ac}/V_t} - 1) \tag{2.1}$$

where v_a is the anode-to-cathode voltage across the diode, I_s is the saturation current, and V_t is the thermal voltage. This equation can be represented by a curve in the two-dimensional plane where the axes are the device anode current and the voltage between anode and cathode. In this case, the cathode node has been chosen as the reference node.

If the device currents can be uniquely determined from the equations, given any value of device voltages, then the device equations are said to be *voltage-controlled.* Often, given a set of device equations, it is possible to perform a transformation so that the device currents are explicit functions of the device voltages.

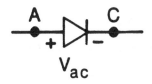

Figure 2.1 - A Diode in Free Space

For example, the diode current equation above is voltage-controlled; the current is an explicit function of the device voltage. Another important example of voltage-controlled equations is the commonly used approximate equations for the MOS transistor shown in Fig. 2.2:

$$i_d = \frac{k}{2} \frac{W}{L} [\, 2(v_{gs} - V_T)v_{ds} - v_{ds}^2 \,] \quad for \ v_{ds} \leq v_{gs} - V_T \qquad [2.2]$$

$$i_d = \frac{k}{2} \frac{W}{L} (v_{gs} - V_T)^2 \quad for \ v_{ds} > v_{gs} - V_T$$

$$i_g = 0$$

where i_d is the drain current, k is a parameter depending on the carrier mobility and on the thickness of the oxide, W and L are the width and the length of the channel of the transistor, v_{gs} is the gate-to-source voltage, v_{ds} is the drain-to-source voltage, V_T is the threshold voltage, and i_g is the gate current.

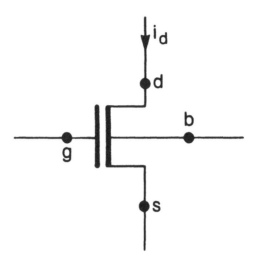

Figure 2.2 - An MOS Transistor in Free Space

The reference terminal in this case is the source terminal. Note that the branch current equations for the MOS transistor are specified by two different algebraic functions, where the function used is determined by the voltages at the terminals.

If the branch terminal voltages can be uniquely computed given the device terminal currents, then the equations are said to be *current-controlled*. Note that most of the devices in use today can be expressed by voltage-controlled equations.

Branch equations for an n-terminal capacitive device are a set of $(n-1)$ algebraic equations involving terminal voltages and terminal charges. Just as in the resistive case, one terminal voltage is taken to be the reference, and one terminal charge can be discarded because the sum of the charges must be a constant, according to Gauss's law[43]. Since absolute charge is unimportant in the context of circuit simulation, most charge models arbitrarily force this constant to be zero. Capacitive device equations can be geometrically represented by a surface in the $2(n-1)$-dimensional space where the axes are all but one of the terminal charges and all the voltages but the reference terminal voltage. If the terminal charges can be uniquely computed from the equations, given any set of terminal voltages, then the device charge equations are said to be *voltage-controlled.* For most commonly modeled devices, there exists a set of equations for the terminal charges that is voltage-controlled. For example, consider the junction capacitance for the diode in Fig. 2.1 for the case where the voltage across the diode $v_{ac} < 0.0$. Then, the anode charge, q_a, can be computed (to first order) with the equation

$$q_a = C_0 (\frac{1 - v_{ac}}{\phi})^{1-m} \qquad [2.3]$$

where C_0 is the zero-bias junction capacitance and ϕ is the junction potential. The charge equations for the capacitors of the MOS transistor shown in Fig. 2.2 are, to a first-order approximation, piecewise linear. For example, in the saturation region we have

$$q_g = \frac{2}{3}WLC_{ox}(v_{gs} - V_T) + C_p(v_{gs} - v_{ds}) \qquad [2.4a]$$

$$q_d = 0 \qquad [2.4b]$$

where q_g and q_d are respectively the charge stored at the gate and the charge stored at the drain of the device, C_{ox} is the oxide capacitance, and C_p is a small parasitic capacitance.

Branch equations for n-terminal inductive devices can be represented by a set of algebraic equations involving $2(n - 1)$ terminal currents and fluxes. If the terminal fluxes can be solved uniquely from the equations given any set of terminal currents, then the device is said to be *current-controlled.* Flux-controlled devices can be defined similarly.

Often, given a physical device, its electrical behavior is best described by a combination of resistive, capacitive and inductive branch equations. An example is the diode of Fig. 2.1, where the resistive equations and the capacitive equations have been introduced above. Often, the branch equations are symbolically represented by *ideal elements* such as two-terminal linear and nonlinear resistors, two-terminal capacitors and inductors, and controlled sources. For example, the model of the MOS transistor introduced above is shown in Fig. 2.3, where the capacitances for the device are represented.

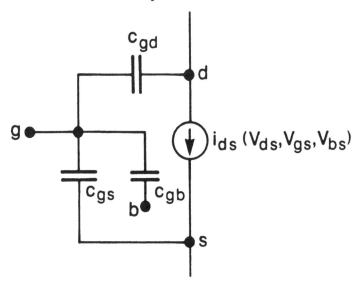

Figure 2.3 – An MOS Transistor Model

SECTION 2.1.2 – KCL and KVL

There are many ways of expressing KVL and KCL equations [36]. We review only the ones that are particularly well suited for simulation.

The node equation formulation is the standard way of expressing KCL in circuit simulation. In this formulation, one equation is generated for each node in the circuit by stating that the algebraic sum of the resistive currents at a node is equal to the rate of change of the algebraic sum of charges at the node. Since any current leaving one node must arrive at another node with opposite sign, and the sum of all the node charges must be zero by Gauss's Law, if a system were constructed using the KCL equations for every node in the circuit, the system would be overdetermined. For this reason, the equation for an arbitrary node in the circuit, referred to as the *reference* or *ground* node, is discarded. Summarizing, for each node *j* in the circuit, except for the reference node, we form the following equation:

$$\frac{d}{dt} \sum_{capacitive\ elements\ at\ j} q_{element} + \sum_{other\ elements\ at\ j} i_{element} = 0. \qquad [2.5]$$

KVL equations are not usually written explicitly, but are implied by expressing the relation between node voltages, the voltages of each of the node of the circuit measured with respect to the reference node of the *circuit,* and branch voltages, the voltages of each terminal of an *n*-terminal device measured with respect to the reference node of the *device.* This relation is that the branch voltage is the difference between the terminal voltage measured with respect to the circuit reference node and the node voltage of the reference node for the *n* -terminal device. For example, for the circuit shown in Fig. 2.4, an MOS *nand* gate,

$$v_{ds2} = v_2 - v_1 \qquad [2.6]$$

where v_2 is the node voltage of the drain of the middle transistor with respect to ground, and v_1 is the node voltage of the source node of the same transistor.
Note that the source node was used as a reference node to express the device equations.

SECTION 2.1.3 - Nodal Analysis

KVL, KCL and branch equations completely characterize the electrical behavior of the circuit. Thus, simply writing all these equations and then solving them is enough to compute all the voltages and currents in the circuit. The sparse tableau analysis method[68] consists of exactly this procedure. However, the system of equations which results is very large and, to solve it efficiently, compli-

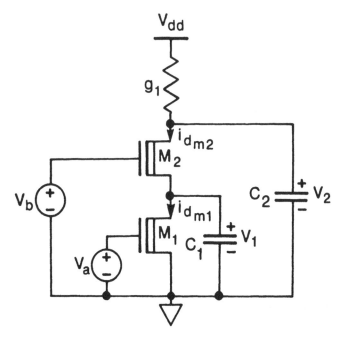

Figure 2.4 - An MOS Nand Gate

cated programming techniques are necessary. Thus, even though it can be shown that this method can yield a simulation which is at least as fast as the one yielded by any other method, other techniques that combine the KVL, KCL and branch equations to yield a smaller, more structured system of equations have been used in a number of successful circuit simulators[2]. In particular, methods derived from the nodal analysis technique share some of the simplicity and efficiency of the sparse tableau without the penalty of complicated programming techniques.

The nodal analysis method can only be applied to circuits where all the devices have voltage-controlled branch equations, and thus is somewhat limited. For example, circuits with inductors and independent voltage sources cannot be analyzed directly with this method. But, as we shall show below, these elements can be included by using very simple extensions of the method.

Nodal analysis consists of the following logical steps. KCL equations are written for all the nodes in the circuit except the ground or reference node. Then, the branch equations are used to express the branch currents in terms of the branch voltages. Note that in this step we need to have voltage-controlled branch equations. Finally, the KVL is applied to express the branch voltages in terms of the node voltages. The result is a set of $(n-1)$ equations which have the node voltages as unkowns. In particular, we have constructed a system of the form

$$\frac{\partial}{\partial t} q(v(t), u(t)) = i(v(t), u(t)) \qquad [2.7]$$

where $q(v(t), u(t)) \in \mathbb{R}^{n-1}$ is the vector of sums of capacitive charges at each node, $i(v(t), u(t)) \in \mathbb{R}^{n-1}$ is the vector of sums of resistive currents at each node, $u(t) \in \mathbb{R}^m$ is the vector of input voltages, and $v(t) \in \mathbb{R}^{n-1}$ is the vector of node voltages.

When independent voltage sources are present in the circuit, the elimination of the branch currents in favor of the branch voltages cannot be accomplished. However, in the simple case where the independent voltage sources are connecting a node of the circuit to the reference node, it is trivial to adapt the method. In fact, the node equation for the node connected to the voltage source is discarded and the respective node voltage is replaced by the value of the source, which is, of course, known.

An example of nodal analysis formulation is given for the MOS *nand* circuit of Fig. 2.4:

$$i_{d_{m1}}(v_1, V_a, 0) - i_{d_{m2}}(v_2, V_b, v_1) + \qquad [2.8a]$$

$$\frac{d}{dt}[\, q_{d_{m1}}(v_1, V_a, 0) + q_{s_{m2}}(v_2, V_b, v_1) + c_1 v_1 \,] = 0$$

and for the second node,

$$i_{d_{m2}}(v_2, V_b, v_1) + g_1(v_2 - V_{dd}) + \frac{d}{dt}[\, q_{s_{m2}}(v_2, V_b, v_1) + c_2 v_2 \,] = 0 \qquad [2.8b]$$

where $i_{d_{m1}}$ and $i_{d_{m2}}$ are the the currents flowing from the drains to the sources of transistor $m1$ and $m2$ respectively, and $q_{d_{m1}}$, $q_{d_{m2}}$, $q_{s_{m2}}$, are the charges accumulated at the drain of transistor $m1$ and the drain and source of transistor $m2$ respectively.

For the purpose of the following analyses, we can summarize the node equations in the following general formulation for a circuit with $n + 1$ nodes. Let q be the vector of the sum of the charges associated with each of the nodes due to capacitive devices, let x be the vector of node voltages of the circuit, then define

$$C(x(t),u(t)) \;=\; \frac{\partial q}{\partial x}(x(t),u(t)) \qquad\qquad [2.9a]$$

and

$$f(x(t),u(t)) \;=\; i(x(t), u(t)) - \frac{\partial q}{\partial u}(x(t),u(t))\dot{u}(t). \qquad\qquad [2.9b]$$

Then, the nodal equations can be expressed as:

$$C(x(t), u(t))\,\dot{x}(t) \;=\; f(x(t), u(t)) \qquad x(0) = x_0 \qquad\qquad [2.10]$$

where $x(t) \in \mathbb{R}^n$ on $t \in [0,T]$; $u(t) \in \mathbb{R}^m$ on $t \in [0,T]$; and $C: \mathbb{R}^n x \mathbb{R}^m \to \mathbb{R}^{nxn}$

SECTION 2.1.4 - Extending the Nodal Analysis Technique

The nodal analysis technique can only be used to form the equations of circuits with elements whose currents or charges are well-behaved functions of voltages. It is possible to extend this technique to include circuits with inductors and floating voltage sources by using modified nodal analysis[38]. A simplification of modified nodal analysis is presented in this section to demonstrate that circuits with inductors and floating voltage sources can be described by a system of equations of the form of Eqn. (2.10). This implies that the form of Eqn. (2.10) can encompass much more that just circuits with voltage-controlled current and charge elements, and is a justification for considering only systems of the form of Eqn. (2.10) for rest of this book.

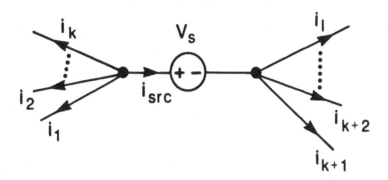

Figure 2.5 - A Floating Voltage Source

Consider a large network with two nodes that are connected by a floating voltage source as in Fig. 2.5. The nodal analysis equations can be written for the two nodes and are for node a,

$$\sum_{j=1}^{k} i_j(v_a, v_b, v) + i_{src} = 0 \qquad [2.11a]$$

and for node b,

$$\sum_{j=k+1}^{l} i_j(v_a, v_b, v) - i_{src} = 0 \qquad [2.11b]$$

where v is the vector of all the other node voltages and i_{src} is the current through the voltage source. Given an additional variable has been introduced, i_{src}, an additional equation is needed to compute the solution,

$$v_a = v_b + V_{src}.$$ [2.11c]

In order to convert this set of equations into the form of Eqn. (2.10) we eliminate i_{src} from the set of equations by adding the two node equations and we substitute one of the node voltages (here we have arbitrarily chosen v_a) with the linear combination of the other and of the voltage of the source.

$$\sum_{j=1}^{k} i_j(v_b + V, v_b, v) = 0.$$ [2.12]

It is somewhat more complicated to reorganize the equations of circuits with inductors so that they fit into the form of Eqn. (2.10). This is because the voltage across the inductor is a function of the time derivative of the current passing through it. For the example in Fig. 2.6a, the KCL equation for node a is

$$\sum_{i=1}^{j} i_i(v_a, v_b, v) + i_{ind} = 0$$ [2.13a]

and for node b,

$$\sum_{i=j+1}^{k} i_i(v_a, v_b, v) - i_{ind} = 0$$ [2.13b]

and for the inductor,

$$L\frac{di_{ind}}{dt} - (v_a - v_b) = 0$$ [2.13c]

where v_a and v_b are the voltages at the inductor terminals, v is the vector of node voltages for the entire circuit excluding v_a and v_b, i_{ind} is the inductor current, and L is its inductance.

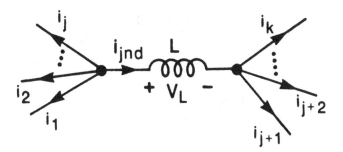

Figure 2.6a - An Inductor Example

We now replace the inductor by an extra circuit node, a grounded capacitor of capacitance L, and two voltage-controlled current sources (See Fig. 2.6b). The circuit equations can now be assembled using nodal analysis as described above. Note that the transformation introduces an additional variable and equation to the system.

SECTION 2.2 - MATHEMATICAL PROPERTIES OF THE EQUATIONS

It has been shown that Eqn. (2.10) is the general form in which the circuit equations can be cast. Before considering techniques for the numerical solution of such equation systems, we will examine some of the mathematical properties that the equation systems inherit from the circuits they describe.

SECTION 2.2.1 - Existence of Solutions

Obviously, since a circuit exists physically, there must be a solution to the problem of how it behaves as time progresses. However, the system of ODE's which describe the circuit contains a variety of implied approximations, and these

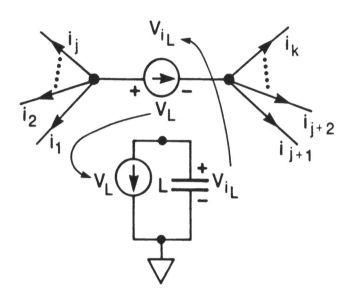

Figure 2.6b - An Inductor Equivalent Circuit

approximations may ignore enough of the physics to produce systems with no solution. It is possible to avoid such modeling pitfalls if it is known under what conditions the ODE system will have a solution.

If we require that there exists a transformation of Eqn. (2.10) to the form

$$\dot{y} = F(y,u) \qquad\qquad [2.14]$$

where u is the vector representing the independent sources and y is a subset of the node voltages, then if u is piecewise continuous with respect to t, and F is Lipschitz continuous with respect to y for all u, i.e., there exists a constant L , called the Lipschitz constant, such that for all pairs y_1, y_2, $\| f(y_1, u) - f(y_2, u) \| \le L \| y_1 - y_2 \|$ for all u, then a unique solution for the system exists[39].

If $C: \mathbb{R}^n x \mathbb{R}^m \rightarrow \mathbb{R}^{nxn}$ is such that $C(x, u)^{-1}$ exists for all x and u, then Eqn. (2.10) can be transformed to the form of Eqn. (2.14). If, in addition, $C(x, u)^{-1}$ is uniformly bounded with respect to x, u, i.e. there exists a constant γ such that

$\|C(x, u)^{-1}\| \le \gamma$ for all x, u, and $f: \mathbb{R}^n x \mathbb{R}^m \to \mathbb{R}^n$ is Lipschitz continuous with respect to x for all $u(t) \in \mathbb{R}^m$, then the function F in the transformed system will be Lipschitz continuous. Finally, if it is also true that $u(t)$ is piecewise continuous, there exists a unique solution to Eqn. (2.10) on any finite interval $[0,T]$ [39].

The fact that $C(x, u)$ has a well-behaved inverse guarantees the existence of a normal form (the form of Eqn. 2.14) for Eqn. (2.10), and that $x(t) \in \mathbb{R}^n$ is the vector of state variables for the system. Note that in order for $C(x, u)$ to have a well-behaved inverse, each node in the circuit must have at least one capacitive device connected to it, otherwise the matrix would have a row of all zeros and hence would be trivially singular. Also, in order for the f defined in Eqn. (2.10) to satisfy the Lipschitz continuity property, the branch equations of the resistive devices have to be Lipschitz continuous, and either $\dfrac{\partial q}{\partial u}$ must be zero, or $\dot{u}$ must be bounded.

SECTION 2.2.2 – Diagonal Dominance and The Capacitance Matrix

As will be shown in subsequent chapters, many of the results concerning relaxation methods applied to systems of the form of Eqn. (2.10) assume that the capacitance matrix has a property referred to as diagonal dominance.

Definition 2.1: A matrix $A \in \mathbb{R}^{nxn}$ is said to be diagonally dominant if for each $i \in \{1,..., n\}$

$$|a_{ii}| \ge \varepsilon + \sum_{j=1, j \ne i}^{n} |a_{ij}| \qquad [2.15]$$

where a_{ij} is the ij^{th} element of the matrix A and ε is a real number such that $\varepsilon \ge 0$. If Eqn. (2.15) hold for each i for some $\varepsilon > 0$, then the matrix is said to be strictly diagonally dominant. ∎

For a broad class of circuits, the $C(x,u)$ matrix defined by $C(x,u) = \dfrac{\partial q}{\partial x}(x,u)$ is strictly diagonally dominant uniformly in x, u; i.e., for any x,u, $C(x,u)$ satisfies Eqn. (2.15) with an $\varepsilon > 0$ that is independent of x,u . Note that the strict diagonal dominance property of $C(x,u)$ guarantees the existence of a bounded inverse[28].

A demonstration of the conditions for which the capacitance matrix will have the strict diagonal dominance property is given by the two-node example in Fig. 2.7. Applying the nodal analysis technique described above yields the following differential equations:

$$(c_1 + c_f)\dot{v}_1(t) - c_f \dot{v}_2(t) = -g_1 v_1(t) \qquad [2.16a]$$

$$(c_2 + c_f)\dot{v}_2(t) - c_f \dot{v}_1(t) = -g_2 v_2(t). \qquad [2.16b]$$

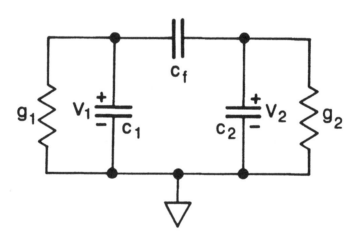

Figure 2.7 - Floating Capacitor Example

As this example demonstrates, for circuits whose only capacitive devices are two-terminal capacitors, the ith diagonal entry of the C matrix is the sum of the capacitance incident at node i, and the ij^{th} entry is the negative value of the capacitance between node i and node j. It therefore follows that in any given row, the sum of the absolute values of the off-diagonal terms is less than or equal to the diagonal term, where the inequality is strict if there is a nonzero capacitance between node i and a voltage source or ground node. This example leads to the following important observation which is easily verified.

Observation If a system of equations of the form of Eqn. (2.10) is constructed by applying the nodal analysis technique described above to a circuit which contains capacitors (linear or nonlinear) or any other elements whose charge function has a diagonally dominant Jacobian, then the capacitance matrix $C(x,u)$ of Eqn. (2.10) is diagonally dominant. If, in addition, there exist a linear or nonlinear capacitor, whose capacitance is bounded away from zero, to ground or a voltage source at each node in the circuit, the matrix $C(x,u)$ is strictly diagonally dominant uniformly in x, u..

SECTION 2.2.3 - Resistor-and-Grounded-Capacitor (RGC) Networks

An important special class of circuits are those constructed from two-terminal positive resistors and two-terminal positive capacitors, where one node of each capacitor is connected to a known voltage (either an independent voltage source or ground). Such circuits occur in practice when modeling the parasitic components in the interconnection networks of circuits, or when using simplified transistor models[67]. In this section we will show that given certain very mild assumptions, such circuits are represented by matrices that are connectedly diagonally dominant[28], a property that we will use in Chapter 3.

Given a matrix $A \in \mathbb{R}^{n \times n}$ and a pair $i,j \in \{1,..., n\}$, i is said to be connected to j in A if there exists a sequence of nonzero matrix entries, a_{ik_1}, $a_{k_1 k_2}$,...., $a_{k_r j}$ where $k_1,..., k_r \in \{1,..., n\}$. Using this notion of connection in a matrix, we can define connected diagonal dominance.

Definition 2.2: A matrix $A \in \mathbb{R}^{n \times n}$ is connectedly diagonally dominant if A is diagonally dominant and for each $i \in \{1,..., n\}$ either

$$|a_{ii}| > \sum_{j=1, j \neq i}^{n} |a_{ij}| \qquad [2.17]$$

or i is connected to some $k \in \{1,..., n\}$ for which Eqn. (2.17) holds. ∎

Note that this definition is a variant of the irreducibly diagonally dominant property defined in the literature[28].

A demonstration of the conditions for which an RGC network will have the connected diagonal dominance property is given by the three-node example in Fig. 2.8. Applying nodal analysis yields the following differential equations:

$$c_1\dot{v}_1(t) = -g_1(v_1(t) - v_{in}(t)) - g_2(v_1(t) - v_2(t)) - g_3(v_1(t) - v_3(t)) \qquad [2.18a]$$

$$c_2\dot{v}_2(t) = -g_2(v_2(t) - v_1(t)) \qquad [2.18b]$$

$$c_3\dot{v}_3(t) = -g_3(v_3(t) - v_1(t)) \qquad [2.18c]$$

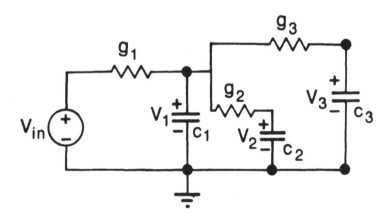

Figure 2.8 – Resistor-and-Grounded-Capacitor Example

The differential equations can be converted to normal form by dividing by the grounded capacitors as follows:

$$\dot{v}_1(t) = -\frac{g_1}{c_1}(v_1(t) - v_{in}(t)) - \frac{g_2}{c_1}(v_1(t) - v_2(t)) - \frac{g_3}{c_1}(v_1(t) - v_3(t)) \qquad [2.19a]$$

$$\dot{v}_2(t) \;=\; -\frac{g_2}{c_2}(v_2(t) - v_1(t)) \qquad\qquad [2.19b]$$

$$\dot{v}_3(t) \;=\; -\frac{g_3}{c_3}(v_3(t) - v_1(t)). \qquad\qquad [2.19c]$$

As this example demonstrates, RGC networks can be represented in the form $\dot{v} = Av + Bu$, where Bu is the vector of currents due to sources, and the ith diagonal entry of A is the negative of the sum of the resistance incident at node i divided by the capacitance at node i, and the ij^{th} entry is the value of the resistance between node i and node j divided by the capacitance at node i. It therefore follows, assuming positive resistors and capacitors, that the sum of the absolute values of the off-diagonal terms is less than or equal to the diagonal terms, and the inequality is strict if there is a nonzero resistance between node i and a voltage source or ground node. In addition, i is connected to j in the matrix A if there is a path of resistors between node i and node j. This example leads to an easily verified observation.

Observation If an RGC network is such that it is possible to trace a path through resistors from any node i to a ground or voltage source node, then the RGC network can be represented in normal form as

$$\dot{v}(t) \;=\; Av(t) + Bu(t) \qquad\qquad [2.20]$$

where $v(t) \in \mathbb{R}^n$ is the vector of node voltages, $u(t) \in \mathbb{R}^m$ is the vector of input sources, $B \in \mathbb{R}^{n \times m}$, and $A \in \mathbb{R}^{n \times n}$ is a connectedly diagonally dominant matrix with negative diagonal entries.

SECTION 2.3 - NUMERICAL INTEGRATION PROPERTIES

Once the system of differential equations has been constructed from the circuit topology, it must be solved numerically. The usual approach is to use one of the many numerical integration formulas to convert the differential equations which describe the system into a sequence of nonlinear algebraic equations.

For example, the most obvious numerical integration formula is the *explicit-Euler* algorithm. Given the initial condition $x(0) = x_0$, it is possible to compute an approximation to $x(h)$, $h > 0$, by substituting $\dfrac{\hat{x}(h) - x(0)}{h}$ for $\dot{x}(0)$, where the notation $\hat{x}$ is used to indicate numerical approximation. Substituting this discrete approximation into Eqn. (2.10) yields the following equation for $\hat{x}(h)$:

$$\hat{x}(h) = x(0) + h\, C(\hat{x}(0), u(0))^{-1} f(x(0), u(0)). \qquad [2.21]$$

By substituting $\hat{x}(h)$ for $x(0)$ in Eqn. (2.26) it is possible to compute $\hat{x}(2h)$, and the process can be repeated to produce a sequence that approximates the exact solution to the differential equation at discrete points in time.

The explicit-Euler algorithm is the simplest of a wide variety of discretization techniques for numerically solving large systems of differential equations. In order to chose a discretization method that will be efficient and accurate for a given class of problems, it is necessary to consider several properties of the integration method with respect to that class. In this section we will consider several of the key aspects of the circuit simulation problem that impact the choice of a numerical method. We will start by presenting the general classical consistency/stability/convergence criteria, both for completeness and as a vehicle for presenting the notation that will be used throughout this book. We will then consider more specific properties of the circuit simulation problem, starting with the well-known issue of stiffness[1]. Following, the properties of *charge con-servation* and *domain of dependence* will be defined, and in each case we will con-sider the impact these properties have on the choice of numerical integration method.

SECTION 2.3.1 - Consistency, Stability, and Convergence

In general, a numerical integration formula produces a sequence approxi-mation to the solution of a differential equation by repeated application, starting from some initial condition x_0. We will denote the approximation produced by the mth application of a given numerical integration formula to Eqn. (2.10) by $\hat{x}(\tau_m)$, where $\tau_m \in \mathbb{R}$ is such that $\hat{x}(\tau_m)$ is the numerical approximation to the exact sol-ution at $t = \tau_m$. It will be assumed that if the differential equation is to be solved numerically on $[0,T]$ and that there exists some finite integer M, such that $\tau_M = T$. In addition, we will refer to $h_m = \tau_m - \tau_{m-1}$ as the mth discretization timestep. Finally, we will denote the entire sequence $x(\tau_m)$, $m \in \{0,..., M\}$ by $\{x(\tau_m)\}$.

If a numerical integration algorithm is to be of any use, it must be possible to arbitrarily accurately approximate the exact solution to the differential equation system uniformly over $[0,T]$ by reducing the discretization timesteps. An integration method with this property is said to be *convergent*, defined formally as follows:

Definition 2.3: Let the discretization timesteps be fixed; that is, $h_m = \dfrac{T}{M}$ for all $m \in \{0,..., M\}$. A numerical integration method is *convergent* with respect to Eqn. (2.10) if the *global* error, defined by

$$\max_{m \le M} \| \hat{x}(\tau_m) - x(\tau_m) \|, \qquad\qquad [2.22]$$

goes to zero as $M \rightarrow \infty$ ∎.

For a numerical integration method to be convergent, it must have two properties. The error made in one timestep must go to zero rapidly as the timestep decreases, and the errors should not grow too rapidly over the timesteps. The error made in one timestep is called the *local truncation error* (LTE).

Definition 2.4: Let $\hat{x}(\tau_m)$ be generated by applying one step of a numerical integration formula to a system of the form of Eqn. (2.10) given the sequence $\{\hat{x}(\tau_{\tilde{m}})\}$, $\tilde{m} < m$ such that $\hat{x}(\tau_{\tilde{m}}) = x(\tau_{\tilde{m}})$. Then the *local truncation error* is defined as $\|\hat{x}(\tau_m) - x(\tau_m)\|$. ∎

The best that one could hope to show for general systems is that the global error for the approximation, $\max_{m \le M} \|\hat{x}(\tau_m) - x(\tau_m)\|$, is a function of the sum of the local truncation errors, $\sum_{m=0}^{M} LTE_m$, where LTE_m is the local truncation error at the mth timestep. Given a fixed interval $[0,T]$, and that $M = \dfrac{T}{h}$, this sum is bounded below by $\dfrac{T}{h}LTE_{min}$ where LTE_{min} is the minimum of the LTE's over all m. If this sum is to go to zero as $h \to 0$, then

$$\lim_{h \to 0} \frac{LTE_m}{h} \to 0$$

for all m. This property is known as *consistency* [1] and is shared by all "reasonable" numerical integration methods.

As an example, it is possible to verify that the explicit-Euler algorithm is *consistent* for systems of the form of Eqn. (2.10), by using a Taylor series expansion about $x(\tau_m)$. That is,

$$x(\tau_{m+1}) = x(\tau_m) + h_{m+1}\dot{x}(\tau_m) + \frac{h_{m+1}^2}{2}\ddot{x}(\tilde{\tau})$$

where $\tilde{\tau} \in [\tau_m, \tau_{m+1}]$. From Eqn. (2.21) we get

$$\hat{x}(\tau_{m+1}) = x(\tau_m) + h_{m+1}C(x(\tau_m), u(\tau_m))^{-1}f(x(\tau_m), u(\tau_m)).$$

Substituting for $\dot{x}$ using the identity

$$C(x(\tau_m), u(\tau_m))^{-1} f(x(\tau_m), u(\tau_m)) = \dot{x}(\tau_m)$$

and then subtracting, we get

$$\hat{x}(\tau_{m+1}) - x(\tau_{m+1}) = \frac{h_{m+1}^2}{2} \ddot{\tilde{x}}(\tilde{\tau}) \qquad [2.23]$$

which verifies consistency.

Consistency is not sufficient to guarantee that a numerical integration method is convergent. Consistency only insures that the local errors are small, but does not indicate anything about how the errors propagate from one timestep to the next. To insure convergence we need to verify that the numerical integration method has a second property, that of stability[1].

Definition 2.5: A numerical integration method applied to Eqn. (2.10) is *stable* if there exists an h_0 and a constant $K < \infty$ such that for any two different initial conditions x_0 and x'_0, and any $h = \dfrac{T}{M} < h_0$,

$$\| \hat{x}(\tau_M) - \hat{x}'(\tau_M) \| < K \| x_0 - x'_0 \| . \blacksquare$$

The explicit-Euler algorithm is *stable*, but the proof is lengthy and well-documented elsewhere[1] so we will not repeat it here.

Not surprisingly, we have the following classical result:

Theorem 2.1: If a numerical integration method is consistent and stable with respect to Eqn. (2.10), then it is convergent with respect to Eqn. (2.10). $\blacksquare$

Several different proofs have been given for this basic result[1].

If an integration method is convergent then when the method is used to compute an approximate solution to a differential equation system, sufficient accuracy can be insured by using timesteps that are small enough. Obviously, it is possible to insure that the timesteps are small enough by using extremely small timesteps, but this is very inefficient. Instead, the integration timesteps are usually controlled by using some check on the discretization error. If in any given step the error becomes too large, the timestep is replaced by a smaller timestep.

Usually, the check on the discretization error is some computed estimate of the local truncation error. For the explicit-Euler algorithm, for example, the exact local truncation error at the mth step is $0.5h_{m+1}^2\ddot{x}(\tilde{\tau})$ where $\tilde{\tau} \in [\tau_m, \tau_{m+1}]$. An estimate of the local truncation error of the mth explicit-Euler step can be computed using the following divided-difference estimate for $\ddot{x}$:

$$\ddot{x}(\tilde{\tau}) \simeq \frac{\dfrac{\hat{x}(\tau_{m+1}) - \hat{x}(\tau_m)}{h_{m+1}} - \dfrac{\hat{x}(\tau_m) - \hat{x}(\tau_{m-1})}{h_m}}{0.5(h_{m+1} + h_m)}. \qquad [2.24]$$

Most of the techniques for estimating local truncation error are only estimates, not bounds. In practice, these types of estimates have proved to be reliable, but there are certain common cases where the estimates are much smaller than the actual error. An example of such a case will be presented in Section 2.3.4.

SECTION 2.3.2 - Stiffness and A–Stability

Consider the example in Fig. 2.9, a resistor-capacitor circuit. The differential equation that describes the circuit can be constructed using nodal analysis:

$$\dot{v}(t) = -100v(t) \qquad v(0) = 1.0 \qquad [2.25]$$

where $v(t) \in \mathbb{R}$ is the node voltage. The exact solution for v is $v(t) = e^{-100t}$. If the interval of interest is [0,1], this is a two-time-scale problem. That is, v changes very rapidly compared to the interval of interest.

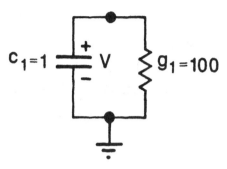

Figure 2.9 – Stiff Resistor-Capacitor Example

Any system of differential equations that has the kind of multiple time-scale properties of the above example is said to be *stiff*. Most circuits of interest generate stiff differential equation systems, and this strongly affects the choice of integration formulas. For example, the explicit-Euler algorithm applied with a fixed timestep h to numerically solve Eqn. (2.25), yields the following recursive equation for $\hat{v}$:

$$\hat{v}(\tau_m) = (1 - 100h_m)\hat{v}(\tau_{m-1}) \qquad [2.26]$$

or given $v(0) = 1$,

$$\hat{v}(\tau_m) = \prod_{i=1}^{m}(1 - 100h_i). \qquad [2.27]$$

Clearly, $|\hat{v}(\tau_m)|$ will decay only if $h_m < 0.02$ for all m, and $\hat{v}(\tau_m)$ will decay monotonically to 0 only if $h_m < 0.01$ for all m. If larger timesteps are used,

$|\hat{v}(\tau_m)|$ will grow. This implies that in order to accurately compute a sequence approximation to the solution of this system using explicit-Euler, small timesteps must be used *even when the solution is not changing appreciably*.

Now consider a slightly different numerical integration formula, the implicit-Euler algorithm, where $\dot{v}(\tau_m)$ is approximated by $\dfrac{1}{h_m}(\hat{v}(\tau_m) - \hat{v}(\tau_{m-1}))$. Just like explicit-Euler, implicit-Euler is convergent, and the local truncation error is of order h^2. When applied to Eqn. (2.27) the following recursion equation results:

$$\hat{v}(\tau_m) = \hat{v}(\tau_{m-1}) - 100\, h_m\, \hat{v}(\tau_m) \qquad [2.28]$$

or reorganizing,

$$\hat{v}(\tau_m) = \frac{1}{(1 + 100h_m)}\hat{v}(\tau_{m-1}). \qquad [2.29]$$

Again using the fact that $v(0) = 1$,

$$\hat{v}(\tau_m) = \prod_{i=1}^{m}(1 + 100h_i)^{-1}.$$

Note that in this case, any $h_m > 0$ will produce a monotonically decaying sequence. The tremendous advantage of this method over explicit-Euler is that small timesteps can be used for the first few steps to accurately resolve the rapid decay, and when the solution stops changing appreciably, the timestep can safely be made orders of magnitude larger without causing the computed solution to grow.

The implicit-Euler algorithm has a property that is "stronger" than the *numerical stability* of Definition 2.5, which we define below as *A-stability*:

Definition 2.6: Let $\{\hat{x}(\tau_m)\}$ be the sequence generated by a numerical integration method applied to the equation

$$\dot{x}(t) = Ax(t) \qquad x(0) = x_0 \qquad\qquad [2.30]$$

where $x(t) \in \mathbb{R}^n$, and $A \in \mathbb{R}^{n \times n}$ and $\tau_m - \tau_{m-1} = h_m = h$ for all m. Given $\{\lambda_i\}$, the set of eigenvalues of A, the *region of stability* for the integration method is the subset of $\mathbb{C}$ such that if $h\lambda_i$ is inside the region of stability for all i, then $x(\tau_m) \to 0$ as $m \to \infty$. The numerical integration method is *A-stable* if the region of stability includes the entire left-half plane of $\mathbb{C}$. ∎

The above definition differs from the original definition given by Dahlquist[42] in that a matrix rather than a scalar test problem is used[6]. As will become apparent in the following sections, a matrix test problem is more appropriate for analyzing methods designed for large systems.

Both the explicit-Euler and implicit-Euler algorithms can be used to produce arbitrarily accurate discrete approximations to the exact solution of Eqn. (2.25), as both are convergent. The implicit-Euler algorithm will allow much larger timesteps to be used with no appreciable loss of accuracy and hence will be more efficient. But improving efficiency is not the only reason one would choose implicit-Euler, or another A-stable numerical integration method. There is also the consideration of numerical robustness. That is, if an A-stable method is used, the timestep can safely be set by considering *only* local truncation error criteria, which can be reasonably estimated. If a method that is not A-stable is used, the timestep must be bounded to insure stability. Such a bound will be a function of the eigenvalues for a linear problem, and it is difficult to get reasonable estimates of eigenvalues.

SECTION 2.3.3 - Charge Conservation

Many differential equation systems generated from physical problems can be characterized by the preservation of certain quantities, and frequently it is important that the numerical method also preserve these quantities. For example, when numerically solving the differential equations that describe the motion of a swinging pendulum in a frictionless environment, it is important to insure that energy remains constant. If energy increases due to numerical error, the computed solution would indicate that the pendulum would swing higher and higher, and if energy were lost, the computed solution would indicate that the pendulum would eventually come to a halt.

In the case of systems of equations that describe circuits, charge is a physical constant. To show this, consider surrounding an arbitrary circuit by a Gaussian surface. Since the surface is unpunctured, the charge contained inside must remain constant[43]. As a consequence, the sum of all the currents must be zero, as the sum of the currents is the derivative with respect to t of the sum of the charge.

This truly trivial observation cannot directly apply to the differential equation systems constructed using nodal analysis as above. If the sum of the node charges in Eqn. (2.7) were precisely zero, then $C(x,u)$ in Eqn. (2.10) would be singular and Eqn. (2.10) would not necessarily have a unique solution. As mentioned above, in order to produce systems of equations that do have unique solutions, the KCL equations for an arbitrary reference node and for nodes for which the voltages are given *a priori* are not included, and a solution for the reference node of $v_{ref}(t) = 0$ for all t is assumed.

As an example, consider the simple resistor-capacitor circuit of Fig. 2.9. In terms of charges, the differential equation that describes the behavior of the circuit is

$$\dot{q}(v(t)) = -gv(t) \qquad v(0) = 1.0, \qquad\qquad [2.31]$$

where the charge $q(v(t)) = cv(t)$. The solution, $v(t) = e^{-\frac{g}{c}t}$, is not a constant, so neither is the charge q. The differential equation does not exhibit charge conservation because not all the charges have been considered, and only the sum remains constant. The charge on the ground node is $-cv(t)$ and obviously the sum of the ground node charge and the charge in Eqn. (2.31) is zero for all t.

If KCL is applied to every node in the resistor-capacitor example, including the reference node, an appended system is generated:

$$\dot{v}(t) - \dot{v}_{ref}(t) = -\frac{g}{c}(v(t) - v_{ref}(t)) \qquad [2.32a]$$

$$\dot{v}_{ref}(t) - \dot{v}(t) = -\frac{g}{c}(v_{ref}(t) - v(t)) \qquad [2.32b]$$

which has an infinite collection of solutions unless it is assumed $v_{ref}(t) = 0$. However, for any of the solutions the sum of the node charges remains constant.

It is possible to use appended systems generated by applying KCL to every node in a circuit to test how well a numerical integration method conserves charge. If the method is applied to the appended system then charge conservation can be checked by summing all the charges at each timestep to insure the sum remains constant. The algebraic equations generated by the numerical integration method can still be solved in the usual fashion, with the known node voltages and a reference voltage used to eliminate the equations associated with the appended differential equations.

Explicit-Euler applied to an autonomous system (independent of $u(t)$) of the form of Eqn. (2.10) constructed from applying KCL to every node in the circuit yields

$$\frac{dq}{dv}(\hat{v}(\tau_m))(\hat{v}(\tau_{m+1}) - \hat{v}(\tau_m)) = h_{m+1} f(\hat{v}(\tau_m)) \qquad [2.33]$$

where $\frac{dq}{dv}(\hat{v}(\tau_m))$ is the necessarily singular Jacobian of $q(\hat{v}(\tau_m))$, the vector of all the node charges. If it is assumed that at τ_m the sum of the node charges $\sum_{i=1}^{n} q_i(\hat{v}(\tau_m)) = K$, where K is some constant, then charge is conserved only if $\sum_{i=1}^{n} q_i(\hat{v}(\tau_{m+1}))$ is also equal to K. This is not necessarily the case, as can be seen from the Taylor series expansion of $q(\hat{v}(\tau_{m+1}))$ about $q(\hat{v}(\tau_m))$,

$$q(\hat{v}(\tau_{m+1})) = q(\hat{v}(\tau_m)) + \frac{dq}{dv}(\hat{v}(\tau_m))(\hat{v}(\tau_{m+1}) - \hat{v}(\tau_m))$$ [2.34]

$$+ \frac{d^2q}{dv^2}(\hat{v}(\hat{\tau}))(\hat{v}(\tau_{m+1}) - \hat{v}(\tau_m))(\hat{v}(\tau_{m+1}) - \hat{v}(\tau_m))$$

where $\hat{v}(\hat{\tau}) \in [\hat{v}(\tau_m), \hat{v}(\tau_{m+1})]$.   Substituting $h_{m+1} f(\hat{v}(\tau_m))$ for
$\frac{dq}{dv}(\hat{v}(\tau_m))(\hat{v}(\tau_{m+1}) - \hat{v}(\tau_m))$ leads to

$$q(\hat{v}(\tau_{m+1})) = q(\hat{v}(\tau_m)) + h_{m+1} f(\hat{v}(\tau_m)) +$$ [2.35]

$$\frac{d^2q}{dv^2}(\hat{v}(\hat{\tau}))(\hat{v}(\tau_{m+1}) - \hat{v}(\tau_m))(\hat{v}(\tau_{m+1}) - \hat{v}(\tau_m)).$$

Summing the node charges,

$$\sum_{i=1}^{n} q_i(\hat{v}(\tau_{m+1})) = \sum_{i=1}^{n} q_i(\hat{v}(\tau_m)) + \sum_{i=1}^{n} h_{m+1} f_i(\hat{v}(\tau_m)) + O(h_{m+1}^2)$$ [2.36]

where $O(\cdot)$ is any function such that $\lim_{\alpha \to 0} \frac{|O(\alpha)|}{\alpha} < \infty$. To simplify Eqn.
(2.36), another property of the original network from which the KCL equations
were generated can be used. Since $f(\cdot)$ is the vector of sums of the currents incident at each node, and as any current leaving a node must arrive at some other
node, $\sum_{i=1}^{n} f_i(\hat{v}(\tau_m))$ must be identically zero. Using this fact leads to

$$\sum_{i=1}^{n} q_i(\hat{v}(\tau_{m+1})) = K + O(h_{m+1}^2),$$ [2.37]

which implies that the sum of the node charges will not remain constant unless
the second-order term in Eqn. (2.10) is zero, which will be true if all the node
charges are linear functions of the node voltages, but will not be true in general.

The sum of the charges is constant in the limit as h_{m+1} goes to zero, so the nonconstant charge can be viewed as another measure of the local truncation error. However, if the same integration method is applied slightly differently, using charge as a state variable, then the sum of the node charges will stay constant regardless of the step-size. To demonstrate this we again apply the explicit-Euler algorithm, but to the system in the form of an autonomous version of Eqn. (2.7). Discretizing the charge function leads to

$$q(v(\tau_{m+1})) - q(v(\tau_m)) = h_{m+1}f(v(\tau_m)).$$ [2.38]

That the sum of charge is constant, independent of the step-size, follows from:

$$\sum_{i=1}^n q_i(v(\tau_{m+1})) = \sum_{i=1}^n q_i(v(\tau_m)) + \sum_{i=1}^n h_{m+1}f_i(v(\tau_m))$$ [2.39]

and the fact, mentioned above, that $\sum_{i=1}^n f_i(\hat{v}(\tau_m)) = 0$.

We use these ideas to precisely define the charge-conservation property.

Definition 2.7: A system of the form of Eqn. (2.10) is of type S if it has the following two properties: for any exact solution the sum $\sum_{i=1}^n q_i(x(t))$ is a constant independent of t; and $\sum_{i=1}^n f_i(v) = 0$ for any $v \in \mathbb{R}^n$. A numerical integration method has the *charge conservation* property if, when applied to any system of type S, the computed sequence $\{\hat{v}(\tau_i)\}$ is such that $\sum_{i=1}^n q_i(v(\tau_m))$ is a constant independent of m. ∎

In Section 3.1 we will show that all multistep integration methods applied with charge as the state variable have the charge conservation property.

SECTION 2.3.4 - Domain of Dependence

In the area of partial differential equations, the concept of *domain of dependence* is well known[44]. The idea is that partial differential equations can be characterized by how rapidly the behavior of points in space will propagate over time. As the time increases, the space of points that can affect a given point, referred to as the given point's domain of dependence, grows. For a numerical method used to solve the partial differential equation to be convergent, that is, to produce arbitrarily accurate solutions as the distance between discretization points becomes small, the numerical method must propagate the behavior of each point in space at a rate that at least approaches the rate of the original partial differential equation. In the language of domain of dependence, a numerical method is convergent only if for each point in space, as the distance between discretization points becomes small the numerical domain of dependence includes, or comes arbitrarily close to covering, the domain of dependence of the partial differential equation.

In this section, an analogous concept will be introduced for large systems of ordinary differential equations. But rather than comparing the domain of dependence of a numerical method to that of the differential equation system to investigate the numerical method's convergence properties, we will show that domain of dependence plays a role in the accuracy of the integration method. In addition, we will discuss to what extent the errors due to discretization can be controlled.

Consider the following differential equation system:

$$\dot{x}_1(t) = -(x_1(t) - 0.01u(t)) \qquad [2.40]$$

$$\dot{x}_2(t) = -(x_2(t) - 10x_1(t))$$

$$\bullet$$

$$\bullet$$

$$\bullet$$

$$\dot{x}_n(t) = -(x_n(t) - 10x_{n-1}(t))$$

$$x_i(0) = 0, \quad i \in \{1,..., n\}$$

where the input $u(t) = 1$ for all $t \geq 0$.

The exact solution for this system is

$$x_i(t) = 10^{i-3}[1 - (\sum_{j=0}^{i-1} \frac{t^j}{j!}) e^{-t}]. \qquad [2.41]$$

As can be seen by examining Eqn. (2.41), the solution to the system of Eqn. (2.40) is a propagating step that is being smoothed and is growing rapidly in amplitude through n stages. Systems with this type of behavior are common among circuit examples (a chain of inverters, for example).

If the explicit-Euler algorithm is applied to Eqn. (2.40), the computed value for $\hat{x}_1(\tau_1) = 0.01h_1$ and $\hat{x}_i(\tau_1) = 0$ for all $1 < i \leq n$. In fact, $\hat{x}_i$ will remain zero until the ith timepoint regardless of the size of the timestep. This slow propagation of the solution introduces an error in the form of a delay; that is, $\hat{x}_i(\tau_j)$ does not change until $j \geq i$. Explicit-Euler is convergent, so this delay error *in time* must go to zero as the timestep decreases, and it does, because τ_j approaches zero. If implicit-Euler is applied to Eqn. (2.40), then $x_i(\tau_1) = \dfrac{10^{i-3}h_1}{(1 + h_1)^i}$. Therefore, when the implicit-Euler algorithm is used, the behavior of the input is propagated throughout the entire system in one timestep and there is no error due to delayed propagation of information. This does *not* necessarily imply that implicit-Euler is more accurate than the explicit-Euler algorithm. For example, applied to Eqn. (2.40) with a timestep $h_1 = 1$, explicit-Euler produces the solution $\hat{x}_5(\tau_1) = 0.0$, while implicit-Euler produces the solution $\hat{x}_5(\tau_1) = 3.125$. The exact solution is $x_5(1) = 0.359$, so in this case, the explicit-Euler computed solution is closer to the exact solution than the implicit-Euler computed solution, though neither method produces very accurate results.

For this example, accuracy clearly is not the reason for preferring the implicit-Euler's rapid propagation of information to explicit-Euler. Implicit-Euler is a more reliable integration method for this example because the error due to discretization in the computed solution is *more visible* than the discretization error in the computed solution produced by the explicit-Euler algorithm. To see why this is the case, consider the local truncation error estimate presented in Section 2.3.1:

$$LTE \simeq h_{m+1}^2 \frac{\dfrac{\hat{x}(\tau_{m+1}) - \hat{x}(\tau_m)}{h_{m+1}} - \dfrac{\hat{x}(\tau_m) - \hat{x}(\tau_{m-1})}{h_m}}{(h_{m+1} + h_m)}. \qquad [2.42]$$

Since in this case, $m = 0, \hat{x}(\tau_m) = \hat{x}(\tau_{m-1}) = x(0)$ and $h_m = 0$, Eqn. (2.42) can be simplified to

$$LTE \simeq h_{m+1}^2 (\hat{x}(\tau_1) - x(0)).$$

For explicit-Euler this estimate indicates that the LTE for $x_5(\tau_1)$ is zero, which is a severe underestimate. A timestep control scheme based on local truncation error would not shrink the timestep in this case, and a very inaccurate solution would be computed. For the implicit-Euler algorithm, the error estimate is 3.125 which is larger than the actual LTE, but this is safe, because an LTE-based timestep control scheme will detect the error and reduce the timestep.

This example indicates that when applying the explicit-Euler algorithm to a large system, a timestep-dependent limit is introduced on how fast the behavior of an individual state variable propagates through the system. The delay error due to this limited rate of propagation is different from a local truncation error. An arbitrarily high-order explicit multistep integration method could have been used at each step, and still $x_i(\tau_m)$ would have been zero until the ith timestep. The implicit-Euler algorithm does not introduce such an *a priori* limitation on how fast the behavior of an individual state variable propagates through the system. Because of this, when the system behavior is faster than can be propagated by the explicit-Euler algorithm, the implicit-Euler algorithm can produce more accurate results, but more importantly, when it produces results that are in error, those errors are more detectable.

We end this section, and this chapter, by connecting the concept of the delay introduced by an integration method, the *numerical delay,* to that of *domain of dependence,* the concept borrowed from the study of partial differential equations. This connection will provide a simple tool for testing integration methods to determine for which types of systems the integration method will introduce numerical delay.

For this purpose, we can define the numerical delay as follows:

Definition 2.8: Given a numerical integration method applied to a system of the form $\dot{x}(t) = Ax(t)$ with some initial condition $x(0) = x_0$, if $(x_i(\tau) - x_i(0)) \neq 0$ for all $\tau \in (0, \tilde{\tau}]$ for some $\tilde{\tau} > 0$, then the *numerical delay to the ith variable* is defined as the smallest integer M_i such that $\hat{x}(\tau_{M_i+1}) - x_i(0) \neq 0$. If no such $\tilde{\tau}$ exists, the numerical delay to the ith variable, M_i, is zero. The *numerical delay* for the integration method applied to the given system with the given initial conditions is the maximum of M_i over all i. ∎

In the example above, the numerical delay for the implicit-Euler algorithm applied to Eqn. (2.40) is zero, and the numerical delay for explicit-Euler is $(n - 1)$.

The description of the role of domain of dependence will be based on the following general definition:

Definition 2.9: Given an equation of the form $y = f(x)$, where $x, y \in \mathbb{R}^n$, and $f: \mathbb{R}^n \to \mathbb{R}^n$, the domain of dependence of the jth variable of the vector y, y_j, is the set of all x_i, $i \in \{1, ..., n\}$ such that for some x, $\dfrac{\partial f_j}{\partial x_i} \neq 0$. ∎

Given the matrix test problem

$$\dot{x}(t) = Ax(t) \qquad x(0) = x_0 \qquad\qquad [2.43]$$

where $x(t) \in \mathbb{R}^n$ and $A \in \mathbb{R}^{n \times n}$, the exact solution at $t = h$ is, in series form,

$$x(h) = [I + hA + \frac{h^2}{2}A^2 + \frac{h^3}{6}A^3 + ...] x(0). \qquad [2.44]$$

The domain of dependence of $x_i(h)$ can be deduced directly from Eqn. (2.44). The variable $x_j(0)$ is in the domain of dependence of $x_i(h)$ if the i, j^{th} element of A^n is nonzero for some n.

The equation for one step of explicit-Euler applied to Eqn. (2.43) is

$$\hat{x}(\tau_1) = [I + h_1 A] x(0). \qquad [2.45]$$

As can be seen from the equation, the domain of dependence for the x_i^{th} variable in Eqn. (2.45) will be a proper subset of the domain of dependence for the x_i^{th} variable in Eqn. (2.44) unless the powers of the matrix A do not add additional nonzero terms. This would occur, for example, in the case where A is diagonal. If instead, one step of implicit-Euler were applied to Eqn. (2.43), the following series expansion results:

$$\hat{x}(\tau_1) = [I + hA + h^2 A^2 + h^3 A^3 + ...] x(0), \qquad [2.46]$$

where the series expansion is valid for h such that $h\rho < 1$ where ρ is the spectral radius of A. Comparing Eqn. (2.46) to Eqn. (2.44), it can be seen that for a small enough h the domains of dependence of the exact solution and the implicit-Euler algorithm are *identical* for each variable, $x_j(h)$. We define this property below as *exhaustive domain of dependence*.

Definition 2.10: If the domain of dependence of each element of the vector produced by one step of an integration method applied to Eqn. (2.43) matches the domain of dependence of the corresponding element in the left-hand-side vector of Eqn. (2.44) for a small enough timestep h and for any A and any initial condi-

tion x_0, then the numerical method is said to have an *exhaustive domain of depend-*
ence. ■

The following theorem relating domain of dependence to numerical delay
follows directly from the definitions:

Theorem 2.2: If a numerical integration method has an exhaustive domain of dependence then the numerical delay of the integration method is zero for any A
and any x_0. ■

If one step of a numerical method has a smaller domain of dependence than
the original differential equation, then a numerical delay will be introduced and
the timesteps used for the calculation will have to be bounded to insure rapid
enough propagation of variable behavior. Like bounds on the timestep to insure
stability for non-A-stable methods, this additional constraint is difficult to estimate, and must be done very conservatively. The explicit-Euler example above
demonstrates how difficult the error is even to observe, because the affected variables, for which the error occurs, are left unperturbed. For this reason, a robust
numerical integration algorithm for large systems must use either a method like
implicit-Euler, which has an exhaustive domain of dependence, or have some
technique for checking that system variables have propagated far enough.

CHAPTER 3 - NUMERICAL TECHNIQUES

As suggested in the previous chapter, in order for a numerical integration method to be reliable and efficient when used to solve the types of differential equation systems generated from circuit descriptions, the method must have several strong properties. For this reason, general purpose circuit simulation programs, like SPICE2[2] and ASTAP[3], use very robust numerical integration algorithms, members of the class of *implicit multistep integration methods*. In the first section of this chapter we will introduce this important class of methods, and demonstrate their use in a simple but nontrivial circuit example. The robustness of the implicit multistep integration algorithms will be demonstrated in Section 2, by proving that the implicit methods used for circuit simulation have the three key properties described in Chapter 2: *charge conservation, exhaustive domain of dependence* and *stiff stability*.

In Section 1 of this chapter it is shown that computing each timestep of an implicit multistep method involves solving an implicit nonlinear algebraic system. Because computing this implicit solution implies solving for all the variables in a given system simultaneously, implicit methods are computationally expensive when applied to large systems. Two approaches have been used to reduce the computation time required by implicit methods. Decomposition techniques have been applied to improve the efficiency of the solution of the large algebraic systems generated by implicit integration algorithms, and less computationally demanding semi-implicit numerical integration algorithms have been developed. In the third section of this chapter the relaxation decomposition algorithms that have been used in circuit simulators for solving the large nonlinear algebraic systems generated by implicit integration methods will be described. In Section 4, the semi-implicit integration methods used in special-purpose programs like MOTIS[7], MOTIS2[8], and SPLICE[45] will be analyzed with respect to their domain of dependence and stability properties. Finally, we will end this chapter by comparing some of the special-purpose integration algorithms with algebraic relaxation methods.

SECTION 3.1 - NUMERICAL INTEGRATION IN GENERAL-PURPOSE SIMULATORS

Most of the general-purpose circuit simulation programs use implicit multi-step integration algorithms applied to the state variable charge (and if inductances are included, also to the fluxes). That is, given a system of the form

$$\dot{q}(x(t), u(t)) = f(x(t), u(t)), \qquad [3.1]$$

where $x(t) \in \mathbb{R}^n$ is the system state, usually the vector of node voltages appended by inductor currents; $u(t) \in \mathbb{R}^l$, is the vector of inputs and is continuously differentiable with respect to t; $f:\mathbb{R}^n x\mathbb{R}^l \to \mathbb{R}^n$ is continuously differentiable and is usually the vector of sums of currents entering a node; and $q:\mathbb{R}^n x\mathbb{R}^l \to \mathbb{R}^n$ is continuously differentiable and is usually the vector of node charges or fluxes. For convenience, a function, $\hat{f}$ is defined such that $\hat{f}(q(x(t)), u(t)) = f(x(t), u(t))$. Using such an $\hat{f}$, Eqn. (3.1) is converted to a system in normal form,

$$\dot{q}(x(t), u(t)) = \hat{f}(q(x(t)), u(t)). \qquad [3.2]$$

One of the collection of multistep integration methods is then used to solve Eqn. (3.2). The general form for a multistep integration method applied to Eqn. (3.2) is

$$\sum_{i=0}^{k} \alpha_i \, q(\hat{x}(\tau_{m-i}), u(\tau_{m-i})) \; = \; h_m \sum_{i=0}^{l} \beta_i \, \hat{f}(q(\hat{x}(\tau_{m-i})), u(\tau_{m-i})) \qquad [3.3]$$

which is identical to

$$\sum_{i=0}^{k} \alpha_i \, q(\hat{x}(\tau_{m-i}), u(\tau_{m-i})) \; = \; h_m \sum_{i=0}^{l} \beta_i f(\hat{x}(\tau_{m-i}), u(\tau_{m-i})) \qquad [3.4]$$

where k,l are positive integers; $\alpha_0 = 1$; and $\alpha_i, \beta_j \in \mathbb{R}$ for $0 < i \le k, 0 \le j \le l$ depend on the integration method and the ratio of the timesteps h_i, $m - \max(k,l) \le i \le m$. For example, the fixed-timestep explicit-Euler algorithm used for examples in Chapter 2 can be derived from Eqn. (3.4) by setting $k = 1$, $l = 1$, $\alpha_0 = 1$, $\alpha_1 = -1$, $\beta_0 = 0$, and $\beta_1 = 1$. To derive implicit-Euler the coefficients remain the same except $\beta_0 = 1$ and $\beta_1 = 0$.

Not all collections of α 's and β's produce useful numerical integration methods. Consistency is one limitation in the choice of coefficients. It is well known that for a multistep method to be consistent,

$$\sum_{i=1}^{k} \alpha_i = -1$$

and

$$\sum_{i=1}^{k} i \, \alpha_i + \sum_{i=0}^{l} \beta_i = 0$$

where it is assumed that $\alpha_0 = 1$[1]. In addition, if $\beta_0 = 0$ the integration method is said to be *explicit* otherwise, the method is *implicit*.

When a multistep method is applied to a system of the form of Eqn. (3.2), the state at the mth step, $\hat{x}(\tau_m)$, is computed by solving

$$q(\hat{x}(\tau_m), u(\tau_m)) - h_m \beta_0 f(\hat{x}(\tau_m), u(\tau_m)) + \qquad [3.5]$$

$$\sum_{i=1}^{k} \alpha_i q(\hat{x}(\tau_{m-i}), u(\tau_{m-i})) - h_m \sum_{i=1}^{l} \beta_i f(\hat{x}(\tau_{m-i}), u(\tau_{m-i})) = 0$$

for $\hat{x}(\tau_m)$ given $\hat{x}(\tau_j)$, $q(\hat{x}(\tau_j), u(\tau_j))$, and $f(\hat{x}(\tau_j), u(\tau_j))$ for all $j < m$.

Implicit nonlinear algebraic systems generated by integration methods are usually solved using the iterative Newton-Raphson(NR) method. The general Newton-Raphson iteration equation to solve $F(x) = 0$ where $x \in \mathbb{R}^n$ and $F:\mathbb{R}^n \rightarrow \mathbb{R}^n$ is

$$J_F(x^k)(x^k - x^{k-1}) = -F(x^{k-1}) \qquad [3.6]$$

where x^{k+1}, x^k are the $(k + 1)^{st}$ and k^{th} Newton iterates respectively, and J_F is the Jacobian of F with respect to x. The iteration is continued until $\|x^k - x^{k-1}\| < \varepsilon$ and $F(x^k)$ is close enough to 0.

The NR algorithm is used for solving the algebraic problems generated by integration methods for several reasons. The NR algorithm is guaranteed to converge if the initial guess is close enough to the exact solution. From this observation it follows that as the exact solution to the differential equation is a continuous function, it is possible to pick a timestep small enough to insure the NR algorithm will converge if the inital guess for the NR algorithm is the previously computed timepoint. Also, the NR algorithm converges quadratically, and will converge independent of the stiffness of the system, both of which follow from the observation that the NR algorithm will solve a linear problem exactly in one step.

If the Newton algorithm is used to solve Eqn. (3.5) for $x(\tau_m)$, the residue at the kth step, $F(x^k(\tau_m))$, is

$$F(x^k(\tau_m)) = q(\hat{x}^k(\tau_m), u(\tau_m)) - h_m\beta_0 f(\hat{x}^k(\tau_m), u(\tau_m)) + \tag{3.7}$$

$$\sum_{i=1}^{k}\alpha_i q(\hat{x}(\tau_{m-i}), u(\tau_{m-i})) - h_m\sum_{i=1}^{l}\beta_i f(\hat{x}(\tau_{m-i}), u(\tau_{m-i})),$$

and the Jacobian $J_F(x^k(\tau_m))$ is

$$J_F(x^k(\tau_m)) = \frac{\partial q}{\partial x}(\hat{x}^k(\tau_m), u(\tau_m)) - h_m\beta_0\frac{\partial f}{\partial x}(\hat{x}^k(\tau_m), u(\tau_m)). \tag{3.8}$$

Then, $\hat{x}^{k+1}(\tau_m)$ is derived from $\hat{x}^k(\tau_m)$ by solving the linear system of equations

$$J_F(\hat{x}^k(\tau_m)) [\hat{x}^{k+1}(\tau_m) - \hat{x}^k(\tau_m)] = -F(\hat{x}^k(\tau_m)). \tag{3.9}$$

The Newton iteration is continued until sufficient convergence is achieved; that is, $\|\hat{x}^{k+1}(\tau_m) - \hat{x}^k(\tau_m)\| < \epsilon$ and $F(\hat{x}^k(\tau_m))$ is close enough to zero.

Note that here, even if the integration algorithm is explicit ($\beta_0 = 0$), Eqn. (3.5) will still be an implicit algebraic problem with respect to $\hat{x}(\tau_m)$. This occurs because the multistep algorithm was applied using charge as a state variable, and charge is a nonlinear function of x.

We now consider the application of these techniques to the example circuit in Figure 3.1, a simplified static memory cell. In this example the memory cell initially contains a "1", that is node 1 is at v_{dd} and node 2 is at 0 volts. The input source, v_{in}, is at zero volts, and this will cause the memory cell to switch states: node 1 will switch to zero volts and node 2 will switch to v_{dd}.

By numerically simulating the circuit while switching states, it is possible not only to verify that the circuit is functionally correct, but also to determine the circuit's performance (in this case, how fast the memory cell switches).

The first step in simulating the circuit is to construct the differential equations that describe it. Referring to Figure 3.1, the equation for node 1 is

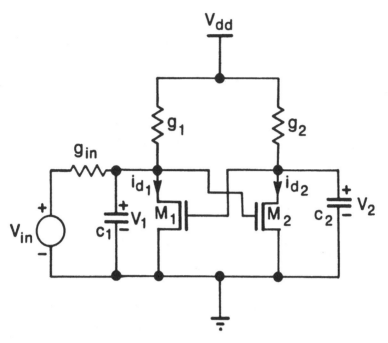

Figure 3.1 - Simplified MOS Static Memory Circuit

$$c_1 \dot{v}_1 + \dot{q}_{d1}(v_1, v_2) + i_{d1}(v_1, v_2) + \hspace{3cm} [3.10a]$$

$$g_1(v_1 - v_{dd}) + \dot{q}_{g2}(v_2, v_1) + g_{in}(v_1 - v_{in}) = 0$$

where i_{d1} is the current from the drain node to the source node of transistor $m1$, q_{d1} is the charge at the drain of transistor $m1$, and q_{g2} is the charge at the gate of transistor $m2$. The equation for node 2 is

$$c_2 \dot{v}_1 + \dot{q}_{d2}(v_2, v_1) + i_{d2}(v_2, v_1) + g_2(v_2 - v_{dd}) + \dot{q}_{g1}(v_1, v_2) = 0 \ [3.10b]$$

where i_{d2} is the current from the drain node to the source node of transistor $m2$, q_{d2} is the charge at the drain of transistor $m2$, and q_{g1} is the charge at the gate of transistor $m1$. The initial conditions are, as mentioned above, $v_1(0) = v_{dd}$ and $v_2(0) = 0$.

For this example system, a fixed-timestep implicit-Euler algorithm will be used to numerically compute the solution. Applying the integration method to

Eqn. (3.10) leads to the following implicit algebraic equation for the voltages at the first timestep h :

$$c_1(v_1(h) - v_1(0)) + q_{d1}(v_1(h), v_2(h)) - q_{d1}(v_1(0), v_2(0)) + \qquad\qquad [3.11a]$$

$$hi_{d1}(v_1(h), v_2(h)) + hg_1(v_1(h) - v_{dd}) +$$

$$q_{g2}(v_2(h), v_1(h)) - q_{g2}(v_1(0), v_2(0)) + hg_{in}(v_1(h) - v_{in}(h)) = 0,$$

$$c_2(v_2(h) - v_2(0)) + q_{d2}(v_2(h), v_1(h)) - q_{d2}(v_2(0), v_1(0)) + \qquad\qquad [3.11b]$$

$$hi_{d2}(v_2(h), v_1(h)) + hg_2(v_2(h) - v_{dd}) + q_{g1}(v_2(h), v_1(h)) - q_{g1}(v_2(0), v_1(0)) = 0.$$

For simplicity, we will ignore the transistor's nonlinear gate and drain charges, and use the following simplified equation for the transistor drain current:

$$i_d(v_{gs}, v_{ds}) \simeq \beta(v_{gs})^2 v_{ds}, \qquad\qquad [3.12]$$

where v_{gs} is the voltage difference between the transistor gate and source nodes, v_{ds} is the voltage difference between the drain and source nodes, and β is a scalar. (βv_{ds} is roughly the transconductance of the MOS transistor in the saturation region (a function of its geometry)). In addition, we will assign the following normalized values for the circuit components: $c_1 = c_2 = 1.0$, $g_1 = g_2 = g_{in} = 1.0$, $\beta = 10.0$ For this example with the above values, the time interval of interest is [0,4], and the solution computed with a fixed timestep of 0.4 is sufficiently accurate.

With the simplified transistor model and the values given above Eqn. (3.11a) and Eqn. (3.11b) become

$$(v_1(0.4) - v_1(0)) + 4(v_1(0.4))^2 v_2(0.4) + \qquad\qquad [3.13a]$$

$$0.4(v_1(0.4) - v_{dd}) + 0.4(v_1(0.4) - v_{in}(0.4)) = 0$$

$$(v_2(0.4) - v_2(0)) + 4(v_2(0.4))^2 v_1(0.4) + 0.4(v_2(0.4) - v_{dd}) = 0. \qquad [3.13b]$$

As mentioned above, the nonlinear algebraic system, Eqn. (3.13a) and Eqn. (3.13b), is solved with the iterative Newton method,

$$\begin{bmatrix} 1.8 + 8v_1^k(0.4)v_2^k(0.4) & 4(v_1^k(0.4))^2 \\ 4(v_2^k(0.4)) & 1.4 + 8v_1^k(0.4)v_2^k(0.4) \end{bmatrix} \begin{bmatrix} v_1^{k+1}(0.4) - v_1^k(0.4) \\ v_2^{k+1}(0.4) - v_2^k(0.4) \end{bmatrix} \qquad [3.14]$$

$$= \begin{bmatrix} -f_1(v_1^k(0.4), v_2^k(0.4)) \\ -f_2(v_1^k(0.4), v_2^k(0.4)) \end{bmatrix}$$

where $v_1^k(0.4)$, $v_2^k(0.4)$ and $v_1^{k+1}(0.4)$, $v_2^{k+1}(0.4)$ are the kth and $(k + 1)^{st}$ Newton iterates; and f_1 and f_2, given by

$$f_1(v_1^k(0.4), v_2^k(0.4)) = (v_1^k(0.4) - v_1(0)) + \qquad [3.15a]$$

$$4(v_1^k(0.4))^2 v_2^k(0.4) + 0.4(v_1^k(0.4) - v_{dd}) + 0.4(v_1^k(0.4) - v_{in}(0.4))$$

and

$$f_2(v_1^k(0.4), v_2^k(0.4)) = (v_2^k(0.4) - v_2(0)) + \qquad [3.15b]$$

$$4(v_2^k(0.4))^2 v_1^k(0.4) + 0.4(v_2^k(0.4) - v_{dd})$$

are the Newton residues. The iteration is continued until $\| v_1^{k+1}(0.4) - v_1^k(0.4) \| < \varepsilon_1$, $\| v_2^{k+1}(0.4) - v_2^k(0.4) \| < \varepsilon_1$, $\| f_1(v_1^k(0.4), v_2^k(0.4)) \| < \varepsilon_2$, and $\| f_2(v_1^k(0.4), v_2^k(0.4)) \| < \varepsilon_2$, where ε_1 and ε_2 are small positive numbers. For this example both were chosen to be 0.001.

The computed results from applying the fixed-timestep implicit-Euler (IE) algorithm over the interval [0,4] is given in column three of Table 3.1. The fourth, fifth, and sixth columns of Table 3.1 are the computed results of integration methods to be discussed in the following sections.

TABLE 3.1 – COMPUTED SOLUTIONS FOR MEMORY CIRCUIT					
STEP	TIME	IE	JSI	SSI	SD
0	0	0	0	0	0
1	0.4	0.210	0.138	0.138	-
2	0.8	0.378	0.222	0.233	0.273
3	1.2	0.553	0.303	0.334	-
4	1.6	0.670	0.386	0.446	0.555
5	2.0	0.802	0.472	0.564	-
6	2.4	0.866	0.560	0.675	0.802
7	2.8	0.905	0.646	0.767	-
8	3.2	0.929	0.724	0.835	0.909
9	3.6	0.942	0.791	0.881	-
10	4.0	0.950	0.842	0.911	0.945

SECTION 3.2 – PROPERTIES OF MULTISTEP INTEGRATION METHODS.

One of the important reasons for applying an integration method to the system in the normal form of Eqn. (3.2) as opposed to the form given in Eqn. (3.1), is that the charge-conservation property of Definition 2.7 holds for any consistent multistep method.

Theorem 3.1: Any consistent multistep method of the form of Eqn. (3.3) has the charge-conservation property. ∎

Proof of Theorem 3.1

Let the system of Eqn. (3.1) be of type S, as given in Definition 2.7. To show charge conservation, the vector elements in Eqn. (3.4) are summed to form

$$\sum_{i=1}^{n}\sum_{j=0}^{k}\alpha_j g_i\, (\hat{x}(\tau_{m-j}), u(\tau_{m-j})) = \sum_{i=1}^{n} h_m \sum_{j=0}^{l}\beta_j f_i(\hat{x}(\tau_{m-j}), u(\tau_{m-j})). \quad [3.16]$$

Interchanging summations yields

$$\sum_{j=0}^{k} \alpha_j \sum_{i=1}^{n} q_i(\hat{x}(\tau_{m-j}), u(\tau_{m-j})) = h_m \sum_{j=0}^{l} \beta_j \sum_{i=1}^{n} f_i(\hat{x}(\tau_{m-j}), u(\tau_{m-j})). \quad [3.17]$$

Since the original system is of type S, $\sum_{i=1}^{n} f_i(\hat{x}(\tau_{m-j}), u(\tau_{m-j})) = 0$. Substituting into Eqn. (3.17) and using $\alpha_0 = 1$,

$$\sum_{i=1}^{n} q_i(\hat{x}(\tau_m), u(\tau_m)) = -[\sum_{j=1}^{k} \alpha_j \sum_{i=1}^{n} q_i(\hat{x}(\tau_{m-j}), u(\tau_{m-j}))]. \quad [3.18]$$

Assuming that charge has been conserved up to the mth step, $\sum_{i=1}^{n} q_i(\hat{x}(\tau_j)) = K$ for $j < m$. Then, as $\sum_{j=1}^{k} \alpha_j = -1$ because the method is assumed consistent,

$$\sum_{i=1}^{n} q_i(\hat{x}(\tau_m), u(\tau_m)) = K, \quad [3.19]$$

which proves the theorem ∎.

Exhaustive domain of dependence is also easy to show for most *implicit* multistep methods.

Theorem 3.2: Any implicit multistep method with $\alpha_1 \neq 0$ has an exhaustive domain of dependence when applied to a system of the form $\dot{x}(t) = Ax(t)$, where $x(t) \in \mathbb{R}^n$, and $A \in \mathbb{R}^{n \times n}$.∎

Proof of Theorem 3.2

The general form for a multistep method applied to $\dot{x}(t) = Ax(t)$ is

$$\sum_{i=0}^{k} \alpha_i \hat{x}(\tau_{m-i}) = h_m \sum_{i=0}^{l} \beta_i A \hat{x}(\tau_{m-i}). \quad [3.20]$$

Reorganizing and using the fact that $\alpha_0 = 1$,

$$\hat{x}(\tau_m) = [I - h_m\beta_0 A]^{-1} [-\sum_{i=1}^{k} \alpha_i \hat{x}(\tau_{m-i}) + h_m\sum_{i=1}^{l}\beta_i A\hat{x}(\tau_{m-i})].$$

Since the method is implicit, $\beta_0 \neq 0$, and for small h_m $[I - h_m\beta_0 A]^{-1}$ can be expanded to yield:

$$\hat{x}(\tau_m) = \qquad\qquad\qquad\qquad\qquad\qquad\qquad\qquad\qquad\qquad [3.21]$$

$$[I + h_m\beta_0 A + (h_m\beta_0 A)^2 + (h_m\beta_0 A)^3 + ...][-\alpha_1 + \beta_1 A]\hat{x}(\tau_{m-1}) +$$

$$[I - h_m\beta_0 A]^{-1} [\sum_{i=2}^{k} -\alpha_i \hat{x}(\tau_{m-i}) + h_m\sum_{i=2}^{l}\beta_i A\hat{x}(\tau_{m-i})].$$

Following the same argument as presented in Section 2.3.4, the variable $\hat{x}_i(\tau_{m-1})$ is in the domain of dependence of $\hat{x}_j(\tau_m)$ if the i,j^{th} element of A^n is nonzero for some n, which matches the differential equation and therefore proves the theorem. ∎

The general question of the region of stability for multistep integration methods has received considerable attention[1,42,46] and the wealth of material on this question will not be reproduced here. Instead, we will mention the results that are most critical for circuit simulation applications. Perhaps the most important result is that there are no A-stable multistep integration methods whose local truncation error is of an order higher than h^3. This is known as the Dahlquist barrier[42]. For this reason, the program SPICE[2] uses a combination of the implicit-Euler algorithm mentioned in Chapter 2 and the trapezoidal rule (corresponding to $\alpha_0 = 1$, $\alpha_1 = -1$, $\beta_0 = 0.5$, $\beta_1 = 0.5$) and, as a user option, can also use the variable-order (up to six) backward-difference methods[1]. The program ASTAP[3] uses the variable-order backward-difference methods. The backward-difference methods of order less than three are A-stable, but the higher-order backward-difference integration methods are only *stiffly stable* that is, the methods' regions of stability include at least the real line in the open left-half plane of $\mathbb{C}$, as well as some sections of the open left-half plane surrounding the real line[1].

SECTION 3.3 - RELAXATION DECOMPOSITION

As mentioned above, the implicit multistep integration methods used in all the general-purpose circuit simulation programs require solving an implicit system of nonlinear algebraic equations at each timestep. The algebraic system is usually cast into the form $F(x) = 0$ where $F: \mathbb{R}^n \rightarrow \mathbb{R}^n$, and $x \in \mathbb{R}^n$, which is then solved using the iterative Newton-Raphson(NR) algorithm as in Eqn. (3.6).

The computation of the Newton iterates can be viewed as two pieces: one is evaluating the function F and its Jacobian J_F, and the other is solving a matrix problem. The computational cost of solving the matrix problem grows super-linearly with the size of the problem, as n^α, where n the number of equations in the system and $\alpha > 1$. Circuit simulation programs are intended to handle large circuits, and as the Jacobian matrices are sparse, sparse matrix techniques[40,44,49] are used to keep α as small as possible. It has been empirically observed that the time to perform a sparse matrix solution grows as $1.2 < \alpha < 1.4$ for the matrices associated with circuit simulation problems. The computational cost of a function evaluation grows linearly with the size of problem, but for circuit simulation problems the evaluation of F and J_F is a complicated task. For each element (transistor, capacitor, resistor, etc.) in the circuit, the currents, the charges and their derivatives must be evaluated. For example, the evaluation of the currents and charges associated with one MOS transistor requires more than a hundred floating-point operations.

Because the computations involved in calculating each transistor's charge and current characteristic are much more complicated than the simpler operations involved in the matrix solution, for small to medium-sized problems the function evaluation time dominates the sparse matrix solution time. It is only when the problem involves more than several thousand equations that the matrix solution time dominates. For this reason, the most useful decomposition techniques applied to circuit simulation problems reduce both the matrix solution time and the function evaluation time.

Two approaches to decomposition have been used in circuit simulation programs. The first, which we will not describe in detail here, is referred to as tearing decomposition. For linear equations, tearing is a form of block LU factorization[4, 5, 47, 48, 49, 50]. Its application to nonlinear systems has led to multi-level Newton algorithms[52]. The second approach, closer to the focus of this book, has been to apply the various forms of the iterative relaxation-Newton or SOR-Newton algorithms[21, 53].

As background for the relaxation-Newton algorithm, we will present an extremely brief description of the Gauss-Jacobi and Gauss-Seidel relaxation methods starting with the algorithms for linear systems. A complete discussion can be found in [28].

The linear problem $Ax - b = 0$ where $x = (x_1,...,x_n)^T$, $b = (b_1,...,b_n)^T$, $x_i, b_i \in \mathbb{R}$, and $A = (a_{ij})$, $A \in \mathbb{R}^{n \times n}$ can be solved exactly using Gaussian elimination (with pivoting) given A is nonsingular. For matrices with certain properties, it is also possible to solve for x in an iterative fashion, where each step of the iteration involves inverting a sequence of one-dimensional problems. For example, there is the Gauss-Jacobi relaxation algorithm presented below.

<u>**Algorithm 3.1** (Gauss-Jacobi Algorithm for solving $Ax - b = 0$)</u>

The superscript k is the iteration count and ϵ is a small positive number.

$k \leftarrow 0$;

Guess some x^0.

repeat {

$$k \leftarrow k + 1$$

foreach ($i \in \{1,..,n\}$)

$$x_i^k = \frac{1}{a_{ii}}[b_i - (\sum_{j=1}^{i-1} a_{ij}x_j^{k-1} + \sum_{j=i+1}^{n} a_{ij}x_j^{k-1})]$$

} until ($\|x^k - x^{k-1}\| \leq \epsilon$)

■

The Gauss-Seidel relaxation algorithm is very similar, and can be generated from Algorithm 3.1 by altering the update equation for x_i^k to

$$x_i^k = \frac{1}{a_{ii}}[b_i - \sum_{j=1}^{i-1} a_{ij}x_j^k + \sum_{j=i+1}^{n} a_{ij}x_j^{k-1}].$$

The Gauss-Jacobi algorithm can be written in matrix form as

$$Dx^k + (L + U)x^{k-1} = b$$

and the Gauss-Seidel algorithm can be written in matrix form as

$$(L + D)x^k + Ux^{k-1} = b$$

where $L,D,U \in \mathbb{R}^{n \times n}$ are strictly lower triangular, diagonal, and strictly upper triangular respectively, and are such that $A = L + D + U$. Taking the difference between the kth and $k - 1^{st}$ iteration we get

$$x^k - x^{k-1} = D^{-1}(L + U)(x^{k-1} - x^k)$$

for Gauss-Jacobi, and

$$x^k - x^{k-1} = (L + D)^{-1}U(x^{k-1} - x^k)$$

for Gauss-Seidel. Note that the iterations are not well defined if D is singular, which will be the case if there is a zero on the main diagonal of A.

It is well known that the Gauss-Jacobi relaxation algorithm will converge independent of the initial guess x_0 if and only if the spectral radius of $D^{-1}(L + U)$ is inside the unit circle. Similarly, the spectral radius of $(L + D)^{-1}U$ must be inside the unit circle for the Gauss-Seidel relaxation algorithm to converge. This will be true, for example, if A is strictly diagonally dominant [28]. Another important property of the relaxation algorithms is their rate of convergence. It can be shown that if the Gauss-Jacobi and the Gauss-Seidel iterations converge, they converge at least linearly. That is, after a sufficiently large number of iterations, the error at each iteration decreases according to:

$$\| x^{k+1} - \tilde{x} \| \leq \varepsilon \| x^k - \tilde{x} \|$$

where $\varepsilon < 1$ and $\tilde{x}$ is such that $A\tilde{x} = b$.

Now consider applying the Gauss-Seidel and Gauss-Jacobi relaxation algorithms to the nonlinear system $F(x) = 0$ where $F(x) = (f_1(x),...,f_n(x))^T$ and $f_i : \mathbb{R}^n \to \mathbb{R}$. At each step of the relaxation, the x_i element is updated by solving the implicit algebraic equation

$$f_i(x_1^k,...,x_{i-1}^k, x_i^{k+1}, x_{i+1}^k,..., x_n^k) = 0 \qquad [3.22a]$$

for Gauss-Jacobi, and

$$f_i(x_1^{k+1},..., x_i^{k+1}, x_{i+1}^k,..., x_n^k) = 0 \qquad [3.22b]$$

for Gauss-Seidel.

It is possible to use the Newton-Raphson algorithm to solve the implicit algebraic systems of Eqn. (3.22a) and Eqn. (3.22b) accurately at each step, but this is not essential. That is, it has been shown that the rate of convergence of the nonlinear relaxation is not reduced if, rather than solving the implicit algebraic systems at each step, only one iteration of the Newton method is used[21]. The algorithms so generated are referred to as the relaxation-Newton methods. The

Gauss-Jacobi-Newton algorithm applied to systems of the form $F(x) = 0$ is (using the notation of Eqn. (3.22a)),

$$x_i^{k+1} = x_i^k - [\frac{\partial f_i(x^k)}{\partial x_i}]^{-1} f_i(x^k), \qquad [3.23a]$$

and the Gauss-Seidel-Newton algorithm is

$$x_i^{k+1} = x_i^k - [\frac{\partial f_i(x^{k+1,i})}{\partial x_i}]^{-1} f_i(x^{k+1,i}) \qquad [3.23b]$$

where $x^{k+1,i} = (x_1^{k+1},..., x_{i-1}^{k+1}, x_i^k,..., x_n^k)^T$.

For example, consider applying the Gauss-Seidel-Newton method to solving the system of equations Eqn. (3.13a) and Eqn. (3.13b). The following two iteration equations then replace the matrix Newton equation, Eqn. (3.14):

$$v_1^{k+1}(0.4) - v_1^k(0.4) = -[1.8 + 8v_1^k(0.4)v_2^k(0.4)]^{-1} f_1(v_1^k(0.4), v_2^k(0.4)) \qquad [3.24a]$$

and

$$v_2^{k+1}(0.4) - v_2^k(0.4) = \qquad [3.24b]$$

$$- [1.4 + 8v_1^{k+1}(0.4)v_2^k(0.4)]^{-1} f_2(v_1^{k+1}(0.4), v_2^k(0.4))$$

where the residue functions f_1 and f_2 are given in Eqn. (3.15a) and Eqn. (3.15b). As before, the iteration is continued until $\| v_1^{k+1}(0.4) - v_1^k(0.4) \| < \varepsilon_1$, $\| v_2^{k+1}(0.4) - v_2^k(0.4) \| < \varepsilon_1$, $\| f_1(v_1^k(0.4), v_2^k(0.4)) \| < \varepsilon_2$, and $\| f_2(v_1^k(0.4), v_2^k(0.4)) \| < \varepsilon_2$.

There is the following general theorem about the convergence of relaxation-Newton methods.

Theorem 3.3: If a given $F:\mathbb{R}^n \rightarrow \mathbb{R}^n$ is continuously differentiable, and if there exists an $x \in \mathbb{R}^n$ such that $F(x) = 0$, then if the Jacobian of F at x, $J_F(x)$, is strictly diagonally dominant there exists some $\delta > 0$ such that both the Gauss-Jacobi-Newton and the Gauss-Seidel-Newton iterations applied to F converge for any x^0 for which $\| x_0 - x \| \leq \delta.$∎

The proof of the above well-known theorem can be found in the references[21]. As a direct consequence, we have the following theorem for the nonlinear algebraic systems generated by consistent multistep integration methods.

Theorem 3.4: Let the Gauss-Seidel-Newton or Gauss-Jacobi-Newton relaxation algorithm be used to solve for $\hat{x}(\tau_m)$ in Eqn. (3.5). If, along with the usual assumptions for existence and uniqueness (Section 2.2.1), $f(x,u)$ is continuously differentiable with respect to x uniformly in u, $\dfrac{\partial q}{\partial x}(x,u)$ is strictly diagonally dominant uniformly over all x and u, and $\hat{x}(\tau_{m-1})$ is used as the starting point for the relaxation, then there exists an $\tilde{h}$ such that for all $h_m \leq \tilde{h}$ the relaxation converges to the solution of Eqn. (3.5). ∎

As an intuitive explanation for why Theorem 3.4 should be true, and why non-convergence should never occur, consider implicit-Euler applied to Eqn. (2.10) with $C(x,u) = C$, where C is strictly diagonally dominant:

$$Cx(\tau_m) = Cx(\tau_{m-1}) + h_m f(x(\tau_m), u(\tau_m)). \qquad [3.25]$$

In the limit as $h_m \rightarrow \infty$, Eqn. (3.25) becomes equivalent to solving $f(x(\tau_m),u(\tau_m)) = 0$ for $x(\tau_m)$ by relaxation. Since little is assumed about f other than Lipschitz continuity, it is unlikely that this problem can be solved, in general, with a relaxation method. However, in the limit as the timestep becomes small, Eqn. (3.25) becomes

$$Cx(\tau_m) = b$$

where $b = Cx(\tau_{m-1})$. This problem can be solved by relaxation because C is strictly diagonally dominant. We formalize this observation in the proof of Theorem 3.4.

Proof of Theorem 3.4:

It is sufficient to show that the system of Eqn. (3.5) satisfies the conditions of Theorem 3.3 for small enough h_m. The Jacobian for the function defined by Eqn. (3.5), J_F, is given in Eqn. (3.8). That J_F is strictly diagonally dominant for a small enough h_m follows directly from the observation that in the limit as $h_m \to 0$, J_F approaches $\dfrac{\partial q}{\partial x}$, which is a strictly diagonally dominant matrix by assumption. Conditions insuring the existence of a continuous (in t) solution have been assumed, and therefore if a consistent multistep method is applied to the problem then in the limit as $h_m \to 0$, $x(\tau_m)$ approaches $x(\tau_{m-1})$.∎

The relaxation-Newton methods have become popular for solving circuit simulation problems for two reasons. The first is that, as mentioned in Chapter 2, for a broad class of circuits the capacitance matrix is diagonally dominant and therefore the relaxation-Newton algorithms are guaranteed to converge if the timestep is made small enough. (Note that relaxation-Newton methods are unlike the standard NR methods in that they introduce a limitation on the integration timestep even when the problem is *linear*. We will return to this issue at the end of this chapter). The second reason for the popularity of the relaxation-Newton methods is that, with proper application, it is possible both to avoid matrix solutions and to reduce the computation involved in function evaluation. As the system Jacobian is sparse, the ith component of the function F defined in Eqn. (3.7), F_i, will be a function of only a few components of the vector x. During the relaxation-Newton process this sparsity can be exploited by noting whether the components of x on which F_i depends have changed significantly, and if none of them have, not re-evaluating F_i. If, in addition, F_i is close enough to 0, x_i^{k+1} will be equal to x_i^k and need not be recomputed.

If implemented as described above, such a partial evaluation scheme involves substantial checking to see if F_i should be re-evaluated. The time required to perform this checking can be more than the savings due to partial function evaluation. To avoid this, practical relaxation-Newton algorithms are implemented using a *selective-trace* technique[33] that simultaneously determines the order in which the relaxation equations are solved and the portion of the function that must be recomputed.

SECTION 3.4 - SEMI-IMPLICIT NUMERICAL INTEGRATION METHODS

Although certain implicit multistep integration methods have all the desirable properties described in Chapter 2, they are computationally expensive when applied to very large systems partly because each timepoint requires a large matrix solution. Semi-implicit integration methods, as the name implies, are constructed to be as implicit as possible without making it necessary to perform standard matrix solutions to compute the timepoints. In this section we will discuss three semi-implicit methods, all of which have been used in circuit simulation applications.

Most of the semi-implicit methods have been applied to nonlinear systems in normal form (that is, circuits with no floating capacitors and a grounded capacitor at each node). The methods are usually derived by using an A-stable integration method, and presuming that solving the generated algebraic equation with one iteration of a relaxation-Newton method is "good enough". As these methods do not carry the relaxation-Newton algorithm to convergence, they do not inherit all the properties of the A-stable integration methods on which they are based. Hence, they are different integration methods whose properties must be analyzed.

In order to simplify the analysis of the properties of semi-implicit methods, we will consider them applied to the following linear test problem:

$$\dot{x}(t) = Ax(t) \qquad x(0) = x_0 \qquad\qquad [3.26]$$

where $x(t) \in \mathbb{R}^n$, and $A \in \mathbb{R}^{n \times n}$. We will assume that all the semi-implicit methods presented below are consistent, as this is trivial to verify, and focus on the properties of these algorithms with respect to domain of dependence and stiff-stability. Note that this test problem is too simple to indicate an integration method's charge-conservation properties, and that issue will not be considered.

The simplest of the semi-implicit methods is the following mixture of explicit-Euler and implicit-Euler[5,7]:

$$x(\tau_m) = x(\tau_{m-1}) + h_m[Dx(\tau_m) + (L + U)x(\tau_{m-1})] \qquad\qquad [3.27]$$

where $L, D, U \in \mathbb{R}^{n \times n}$ are strictly lower triangular, diagonal, and strictly upper triangular respectively, and are such that $A = L + D + U$. Note that this algorithm is identical to solving the algebraic equations generated by implicit-Euler applied to Eqn. (3.26) with one iteration of a Gauss-Jacobi relaxation scheme, and therefore the algorithm is referred to as the Jacobi-semi-implicit method. Solving for $x(\tau_m)$ leads to

$$x(\tau_m) = (I - h_m D)^{-1}[I + h_m(L + U)] x(\tau_{m-1}). \qquad\qquad [3.28]$$

Since $(I - h_m D)$ is diagonal, its inverse, if it exists, can be computed trivially.

We have the following stability result for the Jacobi-semi-implicit method(See [6] for similar results).

Theorem 3.5: If the matrix A in Eqn. (3.26) is either lower or upper triangular, or if A is a connectedly diagonally dominant matrix with negative diagonal entries, then the region of stability for the Jacobi-semi-implicit method contains the open left-half plane of $\mathbb{C}$. ∎

This theorem is of practical value because, as demonstrated in Chapter 2, the systems of differential equations that describe most practical circuits with resistors and grounded capacitors will be of the form of Eqn. (3.26), have negative diagonal entries, and have the connected diagonal dominance property.

We will use the following well-known theorem in the proof of Theorem 3.5.

Gerschgorin Theorem

   Let $A \in \mathbb{R}^{n \times n}$. For each $i \in \{1,..., n\}$ define δ_i by

$$\delta_i = \sum_{j=1,\, j \neq i}^{n} |a_{i,j}|. \qquad [3.29]$$

Then, if λ is an eigenvalue of A there is at least one $i \in \{1,..., n\}$ such that

$$|\lambda - a_{ii}| \leq \delta_i \qquad [3.30]$$

∎.

The following proof due to Varga[28] is included for completeness.

Proof of the Gershgorin Theorem

   Let λ be any eigenvalue of A with some corresponding eigenvector x. Since x cannot be the zero vector as it is an eigenvector, there exists some $i \in \{1,..., n\}$ such that $|x_i| \geq |x_j|$ for all $j \neq i$ and $x_i \neq 0$. By definition,

$$(\lambda - a_{ii})x_i = \sum_{j=1,\, j \neq i}^{n} a_{i,j} x_j. \qquad [3.31]$$

Dividing by x_i and taking absolute values leads to

$$|\lambda - a_{ii}| \leq \sum_{j=1, j\neq i}^{n} |a_{i,j}| \left|\frac{x_j}{x_i}\right|. \qquad [3.32]$$

By assumption, $\left|\dfrac{x_j}{x_i}\right| \leq 1.0$ and therefore

$$|\lambda - a_{ii}| \leq \delta_i, \qquad [3.33]$$

which proves the theorem ■.

Proof of Theorem 3.5:

To prove the theorem it is sufficient to show that the matrix M defined by

$$M = (I - h_m D)^{-1}[I + h_m(L + U)] \qquad [3.34]$$

has a spectral radius $\rho(M) < 1$ if A has its eigenvalues in the open left-half plane of $\mathbb{C}$ and is connectedly diagonally dominant with negative diagonal entries, or is upper or lower triangular. If A is upper or lower triangular, the eigenvalues of A are the diagonal entries, which must be strictly negative by assumption. If A is triangular, M will be triangular, and the eigenvalues of M will be its diagonal entries. The ith diagonal entry of M can be calculated explicitly, and is $\dfrac{1}{1 - h_m a_{ii}}$ which is less than 1 as $h_m a_{ii} < 0$. To prove the theorem for the case where A is connectedly diagonally dominant and has negative diagonal entries, we apply the Gerschgorin theorem. The eigenvalues of M can be bounded by the following expression in terms of the matrix elements of A:

$$\rho(M) \leq \max_i\left\{ \sum_{j=1}^{n} |m_{ij}| \right\} = \max_i\left\{ \frac{\dfrac{1}{h_m} + \sum_{j=1, j\neq i}^{n} |a_{ij}|}{\dfrac{1}{h_m} - a_{ii}} \right\}. \qquad [3.35]$$

If A is strictly diagonally dominant then the bound of Eqn. (3.35) is less than one (because $a_{ii} < 0$ and $|a_{ii}| > \sum_{j=1, j\neq i}^{n} |a_{ij}|$) and the theorem is proved.

If A is not strictly diagonally dominant, but only connectedly diagonally dominant, then the bound on $\rho(M)$ given by Eqn. (3.35) is precisely one. Suppose this is the case, that $\rho(M) = 1$. Then there exists an eigenvalue of M, λ, such that $|\lambda| = 1$. Let x be the eigenvector associated with λ. By definition, if λ is an eigenvalue with associated eigenvector x, then for any $i \in \{1,..., n\}$

$$m_{ii}x_i + \sum_{j=1, j\neq i}^{n} m_{i,j}x_j = \lambda x_i. \qquad [3.36]$$

As x is an eigenvector, it has some nonzero maximum element, that is, there exists some $i \in \{1,..., n\}$ such that $|x_i| \geq |x_j|$ for all $j \neq i$, and $|x_i| > 0$. Dividing through by x_i in Eqn. (3.36), taking absolute values, and then using the fact that $|\lambda| = 1$ leads to

$$|m_{ii}| + \sum_{j=1, j\neq i}^{n} |m_{i,j}| \, |\frac{x_j}{x_i}| \geq 1. \qquad [3.37]$$

From Eqn. (3.35),

$$\sum_{j=1}^{n} |m_{ij}| \leq 1. \qquad [3.38]$$

If the sum of Eqn. (3.38) is strictly less than one, which would be the case if the ith row of A were strictly diagonally dominant, then Eqn. (3.37) could not be true, and therefore $\rho(M)$ must be less than one, proving the theorem. If the sum of Eqn. (3.38) is precisely one, then Eqn. (3.37) could be true only if $|\frac{x_j}{x_i}| = 1$ for all j for which $m_{i,j} \neq 0$. If $|\frac{x_j}{x_i}| = 1$ then $|x_j| > 0$. This implies that Eqn. (3.37) could also be written for any row j for which $m_{i,j} \neq 0$. If, for any of these j

rows, A is strictly diagonally dominant in that row, then once again Eqn. (3.37) can not be true, and therefore $\rho(M) < 1$. It is possible to continue this process, and since A is connectedly diagonally dominant, eventually Eqn. (3.37) will be written for some row k that is a strictly diagonally dominant row of A, and this will finally imply that $\rho(M) < 1$∎ .

Although the stability of the Jacobi-semi-implicit integration method is substantially better than that of the explicit-Euler algorithm used in Chapter 2, particularly for mostly diagonal problems, the domains of dependence are identical. This can be seen by comparing Eqn. (3.28) to Eqn. (2.45). The effect of this limited domain of dependence can be demonstrated by applying the Jacobi-semi-implicit method to solving the memory circuit example in Section 3.1. Using the values for the timestep and components given in Section 3.1, the equations for the first timestep of the Jacobi-semi-implicit algorithm are

$$v_1(0.4) = v_1(0) - 4(v_1(0.4))^2 v_2(0) - \qquad\qquad\qquad\text{[3.39a]}$$

$$0.4(v_1(0.4) - v_{dd}) - 0.4(v_1(0.4) - v_{in}(0.4))$$

and

$$v_2(0.4) = v_2(0) - 4(v_2(0.4))^2 v_1(0) - 0.4(v_2(0.4) - v_{dd}). \qquad\text{[3.39b]}$$

The computed solution using the Jacobi-semi-implicit method (JSI) is given in Table 3.1. As can be seen by comparing the Jacobi-semi-implicit computed solution to the implicit-Euler computed solution, the Jacobi-semi-implicit computed solution indicates a slower switching time.

It is possible to construct semi-implicit integration methods that have larger domains of dependence than the Jacobi-semi-implicit integration method without requiring a matrix solution. In particular, there is the Seidel-semi-implicit method,

$$x(\tau_m) = x(\tau_{m-1}) + h_m[(L + D)x(\tau_m) + Ux(\tau_{m-1})]. \qquad [3.40]$$

Solving for $x(\tau_m)$ leads to

$$x(\tau_m) = (I - h_m(L + D))^{-1}(I + h_m U)x(\tau_{m-1}) \qquad [3.41]$$

where $(I - h_m(L + D))^{-1}$ is easy to compute because the matrix is triangular. The Seidel-semi-implicit method has stability properties that are similar to those of the Jacobi-semi-implicit method.

Theorem 3.6: If the matrix A in Eqn. (3.26) is is either lower or upper triangular, or if A is a connectedly diagonally dominant matrix with negative diagonal entries, then the region of stability for the Seidel-semi-implicit method contains the open left-half plane of $\mathbb{C}$.∎

In the case that A is connectedly diagonally dominant with negative diagonal entries, Theorem 3.6 follows from arguments similar to those used to prove Theorem 3.5. If A is lower triangular, the Seidel-semi-implicit algorithm is identical to implicit-Euler which is A-stable, and if A is upper triangular the algorithm is identical to the Jacobi-semi-implicit algorithm.

The Seidel-semi-implicit method does not have obviously better stability properties than the Jacobi-semi-implicit method, but it has the clear advantage of a larger domain of dependence. To see this, consider the expansion of $(I - h_m(L + D))^{-1}$ in Eqn. (3.41) for small h_m:

$$x(\tau_m) = \qquad\qquad\qquad\qquad\qquad\qquad\qquad\qquad\qquad\qquad [3.42]$$

$$[I + h_m(L + D) + h_m^2(L + D)^2 + h_m^3(L + D)^3 + ...][I + h_m U]x(\tau_{m-1}).$$

If A is lower triangular, the domain of dependence of the Seidel-semi-implicit method is exhaustive. As long as the lower-triangular portion of A is nonzero, the domain of dependence of the Seidel-semi-implicit method will be larger than that of the Jacobi-semi-implicit method. The effect of this increased domain of dependence can be demonstrated by applying the Seidel-semi-implicit method to solving the memory circuit example in Section 3.1. Using the values for the timestep and components given in Section 3.1, the equations for the first timestep of the Seidel-semi-implicit algorithm are

$$v_1(0.4) = v_1(0) - 4(v_1(0.4))^2 v_2(0) - \qquad\qquad\qquad [3.43a]$$

$$0.4(v_1(0.4) - v_{dd}) - 0.4(v_1(0.4) - v_{in}(0.4))$$

and

$$v_2(0.4) = v_2(0) - 4(v_2(0.4))^2 v_1(0.4) - 0.4(v_2(0.4) - v_{dd}). \qquad [3.43b]$$

The computed solution using the Seidel-semi-implicit method (SSI) is given in Table 3.1. As can be seen by comparing the Seidel-semi-implicit, Jacobi-semi-implicit, and implicit-Euler computed solutions, the Seidel-semi-implicit method does not introduce as much delay as the Jacobi-semi-implicit method, but it still introduces delay compared to the implicit-Euler method.

The Seidel-semi-implicit method includes the domain of dependence due to arbitrarily high powers of the lower-triangular portion of A. The next semi-implicit method we will consider, the symmetric-displacement algorithm[54,6], also includes the domain of dependence due to arbitrarily high powers of the upper-triangular portion of A. Applied to Eqn. (3.26), the symmetric-displacement algorithm is the following two-step process:

$$x(\tau_{m+1/2}) = x(\tau_m) + 0.25h_m[(2L + D)x(\tau_{m+1/2}) + (D + 2U)x(\tau_{m-1})] \quad [3.44a]$$

$$x(\tau_m) = x(\tau_{m+1/2}) + 0.25h_m[(2U + D)x(\tau_{m+1/2}) + (D + 2L)x(\tau_m)].   [3.44b]$$

Note that if A is diagonal, the symmetric displacement algorithm is precisely the trapezoidal rule.

The symmetric-displacement algorithm has several important properties. The local truncation error is of the order h^3, unlike the other semi-implicit methods, whose error is of order h^2[6]. In addition, it has the stability properties given in the following theorem.

Theorem 3.7: If the matrix A in Eqn. (3.26) is lower or upper triangular, or if A is symmetric, or if A has negative diagonal entries and is strictly diagonally dominant, then the region of stability for the symmetric-displacement method contains the open left-half plane of $\mathbb{C}$.∎

The proof of Theorem 3.7 for the case where A has negative diagonal terms and is strictly diagonally dominant follows from the same reasoning as used in the proof of Theorem 3.5. The proof for the case where A is symmetric can be found in [6].

As indicated by Theorem 3.7, for symmetric problems the stability properties of the symmetric-displacement algorithm can be better than those of the Seidel-semi-implicit method. The symmetric-displacement algorithm is always superior to the Seidel-semi-implicit method in two important aspects: its local truncation is of a higher order, and it has a larger domain of dependence for problems that are not lower triangular. To show this, Eqn. (3.44a) and Eqn. (3.44b) are reorganized as

$$x(\tau_m) = [I - 0.25h_m(D + 2L)]^{-1} [I + 0.25h_m(D + 2U)]   [3.45]$$

$$[I - 0.25h_m(D + 2U)]^{-1} [I + 0.25h_m(D + 2L)] x(\tau_m).$$

The expansion of $[I - 0.25h_m(D + 2L)]^{-1}$ will include all the powers of L, and the expansion of $[I - 0.25h_m(D + 2U)]^{-1}$ will include all the powers of U. Note that this does not mean that the symmetric-displacement algorithm has an exhaustive domain of dependence for the same reason that A^2 is not necessarily the same as $L^2 + U^2$: there are possibly the additional cross-product terms LU and UL.

The effect of this increased domain of dependence can be demonstrated by applying the symmetric-displacement algorithm to solving the memory circuit example in Section 3.1. Using the values for the timestep and components given in Section 3.1, the equations for the first timestep of the symmetric-displacement algorithm are

$$v_1(0.4) = v_1(0) - 2(v_1(0.4))^2 v_2(0) + \qquad\qquad [3.46a]$$

$$0.2(v_1(0.4) - v_{dd}) + 0.2(v_1(0.4) - v_{in}(0.4))$$

$$+ 2(v_1(0))^2 v_2(0) + 0.2(v_1(0) - v_{dd}) + 0.2(v_1(0) - v_{in}(0))$$

and

$$v_2(0.4) = v_2(0) - 2(v_2(0.4))^2 v_1(0.4) + 0.2(v_2(0.4) - v_{dd}) \qquad\qquad [3.46b]$$

$$+ 2(v_2(0))^2 v_1(0) + 0.2(v_2(0) - v_{dd}).$$

The second timestep reverses the order,

$$v_2(0.8) = v_2(0.4) - 2(v_2(0.8))^2 v_1(0.4) + 0.2(v_2(0.8) - v_{dd}) + \qquad\qquad [3.47a]$$

$$+ 2(v_2(0.4))^2 v_1(0.4) + 0.2(v_2(0.4) - v_{dd})$$

$$v_1(0.8) = v_1(0.4) - 2(v_1(0.8))^2 v_2(0.8) + 0.2(v_1(0.8) - v_{dd}) + \qquad\qquad [3.47b]$$

$$0.2(v_1(0.8) - v_{in}(0.8)) + 2(v_1(0.4))^2 v_2(0.4) +$$

$$0.2(v_1(0.4) - v_{dd}) + 0.2(v_1(0.4) - v_{in}(0.4)).$$

The computed solution using the symmetric-displacement method (SD) is given in Table 3.1. Notice that there are values for the symmetric-displacement method only for every other step. Since it uses two half-steps, in order to compare it fairly with other methods, the half-step was set equal to the step-size for the other methods. As can be seen by comparing symmetric-displacement with the other semi-implicit methods, the symmetric-displacement algorithm produces the solution closest to implicit-Euler, but it too introduces a delay.

None of the semi-implicit methods mentioned above match the stiffly-stable implicit multistep method for either stability or domain of dependence. However, they have proved to be useful for a variety of circuit simulation applications where either the problem is not too stiff, or is of a mostly diagonal or lower-triangular form. For this reason, extensions of the semi-implicit methods mentioned above to the case where $C(x,u)$ is not diagonal have been pursued[55,6]. Similar results about regions of stability for these extensions have been shown.

SECTION 3.5 - RELAXATION VERSUS SEMI-IMPLICIT INTEGRATION

The relaxation-Newton algorithms described in Section 2 present a bound on the numerical integration timestep to insure that the relaxation converges. This bound is similar to the bound on the timestep to insure stability of the semi-implicit numerical integration methods. In order to demonstrate briefly the similarities of the two approaches, we will end this chapter by comparing the Jacobi-relaxation algorithm applied to solving the implicit-Euler equation, and the Jacobi-semi-implicit integration algorithm. In order to keep the analysis simple, we will use the test problem of Eqn. (3.26)

The timepoint update equation for the Jacobi-semi-implicit algorithm is

$$x(\tau_m) = (I - h_m D)^{-1}[I + h_m(L + U)]x(\tau_{m-1}). \qquad [3.48]$$

The iteration update equation of the Jacobi relaxation applied to implicit-Euler is

$$x^{k+1}(\tau_m) - x^k(\tau_m) = (I - h_m D)^{-1}[h_m(L + U)][x^k(\tau_m) - x^{k-1}(\tau_m)]. \qquad [3.49]$$

The semi-implicit method will be stable if

$$\rho[(I - h_m D)^{-1}(I + h_m(L + U))] < 1 \qquad [3.50]$$

and the relaxation will converge if

$$\rho[(I - h_m D)^{-1}(h_m(L + U))] < 1. \qquad [3.51]$$

The spectral radii of Eqn. (3.50) and Eqn. (3.51) will both be less than 1 for any h_m if A is connectedly diagonally dominant and has negative diagonal entries. The relaxation algorithm is slightly better than the semi-implicit integration method because the spectral radii of Eqn. (3.51) will be less than 1 for any h_m for any diagonally dominant matrix. If A is not diagonally dominant, but has negative diagonal elements, the method that will allow the larger timestep will depend on the signs and magnitudes of the lower-triangular and upper-triangular portions of A.

Although the size of the largest allowable timestep does not conclusively favor semi-implicit integration methods or relaxation methods, relaxation methods are clearly superior with respect to the relative domains of dependence. By carrying the relaxation iteration to convergence, it is assured that the information at a given timestep has propagated "far enough". Therefore, relaxation methods have the exhaustive domain of dependence property, and, as described above, the semi-implicit methods do not.

CHAPTER 4 - WAVEFORM RELAXATION

The multistep numerical integration algorithms for solving ODE systems can become inefficient for large systems where different state variables are changing at very different rates. This is because the direct application of the integration method forces every differential equation in the system to be discretized identically, and this discretization must be fine enough so that the fastest-changing state variable in the system is accurately represented. If it were possible to pick different discretization points, or timesteps, for each differential equation in the system so that each could use the largest timestep that would accurately reflect the behavior of its associated state variable, then the efficiency of the simulation would be greatly improved. This is referred to as the multirate problem[1], and numerical integration methods that allow for different state variables to use different timesteps are called multirate integration methods.

The selective-trace technique for improving the efficiency of relaxation-Newton methods (Section 3.3) can be thought of as a limited multirate integration method. If, at a given timestep, the x_i variable is at its equilibrium (or stationary) point, and the x_j variables on which x_i depends do not change, then x_i will retain the value it had before the timestep. In fact, x_i will never be recomputed until some x_j on which it depends changes. If x_i is bypassed for several timesteps the effect is the same as if a large timestep were used to compute x_i. Therefore, a selective-trace algorithm exploits the kind of multirate behavior that stems from a system in which most of of the variables remain at equilibrium. The selective-trace algorithm cannot, however, exploit of a system for which the state variables have different rates of motion but are not at equilibrium.

Techniques based on semi-implicit integration algorithms have been used both to achieve the kind of limited multirate integration described above, and to achieve full multirate integration methods[4,57]. However, as pointed out in Section 3.4, the semi-implicit integration algorithms do not have all of the properties that make a numerical method for circuit simulation robust. A different approach is to decompose the differential equations in some way *before* introducing discrete approximations. If the differential equations are solved independently, the numerical integration method used for each system can pick its own timestep, thereby achieving full multirate integration. In addition, since *any* numerical integration algorithm (with some restrictions, see Chapter 6) can be used to solve the decomposed systems, it is possible to choose one that retains all the desirable numerical properties described in Chapter 2.

One method for decomposing differential equations is the family of waveform relaxation(WR) algorithms[11]. WR algorithms have captured considerable attention due to their favorable numerical properties and to the success in applying the WR algorithms to the solution of MOS digital circuits. In this chapter the theoretical basis for the WR algorithm will be presented. Waveform relaxation will be introduced with a simple example, which will be followed by the general algorithm applied to systems of the form of Eqn. (2.10). A new proof of the convergence, one that demonstrates that the WR algorithm is a contraction mapping in a particular norm, is then presented. In the third section, the waveform relaxation-Newton(WRN) algorithm, which is the extension to nonlinear differential equations of the relaxation-Newton algorithm given in Section 3.3, is derived. In the final section of this chapter nonstationary WR algorithms are introduced, and it is proved that nonstationary algorithms converge as a direct consequence of the fact that the WR algorithm is a contraction mapping.

SECTION 4.1 - THE BASIC WR ALGORITHM

Consider the two-inverter chain example in Fig. 4.1. The differential equations for the circuit can be written by inspection, and are

$$C_1 \dot{v}_1(t) + C_f(\dot{v}_1(t) - \dot{v}_2(t)) + g_1(v_1(t) - v_{dd}) + i_{d1}(v_1(t), v_{in}) = 0 \qquad [4.1a]$$

and

$$C_2 \dot{v}_2(t) + C_f(\dot{v}_2(t) - \dot{v}_1(t)) + g_2(v_2(t) - v_{dd}) + i_{d2}(v_2(t), v_1(t)) = 0. \qquad [4.1b]$$

where i_{d1} and i_{d2} are the drain currents through transistor m_1 and m_2 respectively, the initial conditions are given as $v_1(0) = v_{10}$ and $v_2(t) = v_{20}$, and the interval of interest is $t \in [0,T]$.

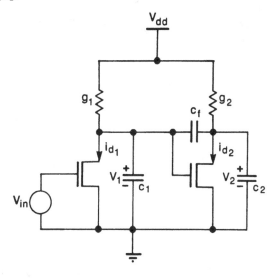

Figure 4.1 - Two-Inverter Circuit

Although Eqn. (4.1a) and Eqn. (4.1b) are coupled, if C_f is small compared to C_1 then v_1 is only weakly coupled to v_2. If this is the case, it is possible to compute a good approximation for $v_1(t)$, $t \in [0,T]$ by solving Eqn. (4.1a) alone, assuming $v_2(t) = v_{20}$ for all $t \in [0,T]$. This approximation to $v_1(t)$, $t \in [0,T]$ can be used as an input *waveform* for Eqn. (4.1b), which can then be solved for $v_2(t)$. This process is demonstrated pictorially in Fig. 4.2. Note that to reverse this process, that is, to solve Eqn. (4.1b) assuming $v_1(t) = v_{10}$ for all $t \in [0,T]$, would not produce a good approximation to the exact solution for $v_2(t)$. This is because $v_2(t)$ is a strong function of the drain current through transistor m_2, which is a strong function of $v_1(t)$.

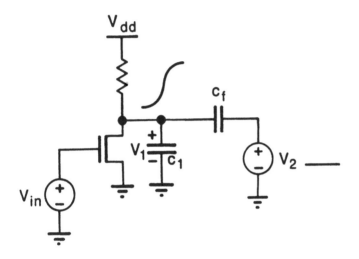

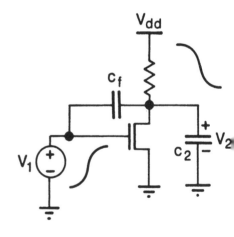

Figure 4.2 - Decomposed Solution Process

In general, MOS digital circuits have this kind of *unidirectional* coupling. That is, it is possible to break the equations into blocks so that if the blocks are solved in the proper order, a good approximate solution to the entire system is produced. Several simulation programs intended to perform approximate analysis of MOS circuits are based on this technique[8,52].

A key, and perhaps obvious, observation is that this kind of single sweep through the differential equation system is like one iteration of a Gauss-Seidel relaxation algorithm applied directly to the differential equations. If one such iteration produces a good approximate solution, then a second iteration will almost certainly produce a more accurate approximate solution, and subsequent iterations will converge rapidly to the exact solution (in the next section we will prove that the iteration is guaranteed to converge). Applying this idea to Eqn. (4.1a) and Eqn. (4.1b) leads to the following iteration equations:

$$C_1 \dot{v}_1^{k+1}(t) + C_f(\dot{v}_1^{k+1}(t) - \dot{v}_2^k(t)) + g_1(v_1^{k+1}(t) - v_{dd}) + i_{d1}(v_1^{k+1}(t), v_{in}) = 0$$

and

$$C_2 \dot{v}_2^{k+1}(t) + C_f(\dot{v}_2^{k+1}(t) - \dot{v}_1^{k+1}(t)) + g_2(v_2^{k+1}(t) - v_{dd}) +$$

$$i_{d2}(v_2^{k+1}(t), v_1^{k+1}(t)) = 0.$$

where the iteration is continued until some convergence criterion of the *waveform* is met. For example, the iteration might be continued until $\max_{t \in [0,T]} \| v_1^{k+1}(t) - v_1^k(t) \| < \varepsilon$ and $\max_{t \in [0,T]} \| v_2^{k+1}(t) - v_2^k(t) \| < \varepsilon$ for some small positive number ε.

As described above, this *waveform relaxation* algorithm can been seen as the analog of the Gauss-Seidel technique for solving nonlinear algebraic equations. Here, however, the unknowns are waveforms (elements of a function space), rather than real variables. In this sense, the algorithm is a technique for the time-domain decoupling of differential equations.

The WR algorithm for solving systems of the form of Eqn. (2.10) is given in Algorithm 4.1. Note that the differential equation in Algorithm 4.1 has only one unknown variable x_i^k. The variables $x_{i+1}^{k-1}, ..., x_n^{k-1}$ are known from the previous iteration and the variables $x_1^k, ..., x_{i-1}^k$ have already been computed. Also, a Gauss-Jacobi version of the WR Algorithm for Eqn. (2.10) can be obtained from Algorithm 4.1 by replacing the **foreach** statement with the **forall** statement and adjusting the iteration indices.

Algorithm 4.1 (WR Gauss-Seidel Algorithm for solving Eqn. (2.10))
The superscript k denotes the iteration count, the subscript i denotes the component index of a vector, and ε is a small positive number.
$k \leftarrow 0$
Guess waveform $x^0(t)$; $t \in [0,T]$ such that $x^0(0) = x_0$
       for example, set $x^0(t) = x_0$, $t \in [0,T]$);
repeat {
    $k \leftarrow k + 1$
    foreach ($i \in \{ 1,..,n \}$) {

      solve

$$\sum_{j=1}^{i} C_{ij}(x_1^k, ..., x_i^k, x_{i+1}^{k-1}, ..., x_n^{k-1}, u)\dot{x}_j^k +$$

$$\sum_{j=i+1}^{n} C_{ij}(x_1^k, ..., x_i^k, x_{i+1}^{k-1}, ..., x_n^{k-1}, u)\dot{x}_j^{k-1} -$$

$$f_i(x_1^k, ..., x_i^k, x_{i+1}^{k-1}, ..., x_n^{k-1}, u) = 0$$

      for ($x_i^k(t)$; $t \in [0,T]$), with the initial condition $x_i^k(0) = x_{i_0}$.

    }
} **until** ($\| x^k - x^{k-1} \| \leq \varepsilon$)
■

SECTION 4.2 - CONVERGENCE PROOF FOR THE BASIC WR ALGORITHM

If the matrix $C(x,u)$ is diagonally dominant and Lipschitz continuous with respect to x for all u then both the Gauss-Seidel and the Gauss-Jacobi versions of

Algorithm 4.1 are guaranteed to converge. It has been shown that the WR algorithm converges when applied to Eqn. (2.10) if $C(x,u)$ is diagonally dominant and
independent of x [12]. In many systems that are modeled in the form of Eqn.
(2.10) C depends on x, so we will present a more general convergence proof that
extends the original theorem to include these systems. In addition, we will prove
the WR algorithm is a contraction in a simpler norm than the one used in the original theorem.

We will prove this WR convergence theorem by first showing that if $C(x,u)$
is diagonally dominant, then there exists a bound on the $\dot{x}^k$'s generated by the
WR algorithm that is independent of k. Using this bound, we will show that the
Lipschitz continuity of $C(x,u)$ implies that there exists a norm on $\mathbb{R}^n$ such that for
arbitrary positive integers j and k,

$$\| \dot{x}^{k+1}(t) - \dot{x}^{j+1}(t) \| \leq$$

$$\gamma \| \dot{x}^k(t) - \dot{x}^j(t) \| + l_1 \| x^{k+1}(t) - x^{j+1}(t) \| + l_2 \| x^k(t) - x^j(t) \|$$

for some $\gamma < 1$ and $l_1, l_2 < \infty$ for all $t \in [0,T]$. In the properly chosen norm
$\| \cdot \|_b$ on $C([0,T], \mathbb{R}^n)$ the above equation implies that

$$\| \dot{x}^{k+1} - \dot{x}^{j+1} \|_b \leq \hat{\gamma} \| \dot{x}^k - \dot{x}^j \|_b$$

where $\hat{\gamma} < 1$ and therefore the sequence $\{\dot{x}^k\}$ converges by the contraction mapping theorem. As $x^k(0) = x_0$ for all k, $\{x^k\}$ converges as well.

Before formally proving this basic WR convergence theorem we will state
the well-known contraction mapping theorem[35], and a few lemmas which will
be used in the course of the proof.

<u>The Contraction Mapping Theorem:</u> Let Y be a Banach space and $F{:}Y \to Y$. If F is
such that $\| F(y) - F(x) \| \leq \gamma \| y - x \|$ for all $x, y \in Y$ for some $\gamma \in [0,1)$, then F
has a unique fixed point $\tilde{y}$ such that $F(\tilde{y}) = \tilde{y}$. Furthermore, for any initial

guess $y^0 \in Y$ the sequence $\{y^k \in Y\}$ generated by the fixed-point algorithm $y^k = F(y^{k-1})$ converges uniformly to $\tilde{y}$ ∎.

Lemma 4.1: If $C(x,u) \in \mathbb{R}^{n \times n}$ is diagonally dominant uniformly over all $x \in \mathbb{R}^n$, $u \in \mathbb{R}^r$ then given any collection of vectors $\{x^1, ..., x^n\}$, $x^i \in \mathbb{R}^n$, and any $u \in \mathbb{R}^r$, the matrix $C^p(x^1, ..., x^n, u) \in \mathbb{R}^{n \times n}$ defined by $C^p_{ij}(x^1, ..., x^n, u) \equiv C_{ij}(x^i, u)$ is also diagonally dominant. In other words, let C^p be the matrix constructed by setting the ith row of C^p equal to the ith row of the given matrix $C(x^i, u)$. Then this new matrix is also diagonally dominant. ∎

Lemma 4.1 follows directly from the definition of diagonal dominance.

Lemma 4.2: Let $C \in \mathbb{R}^{n \times n}$ be any strictly diagonally dominant matrix. Let L be strictly lower triangular, U be strictly upper triangular, and D be diagonal, such that $C = L + D + U$. Then $\|D^{-1}(L + U)\|_\infty < 1$ and $\|(D + L)^{-1}U\|_\infty < 1$. ∎

Lemma 4.2 is a standard result in matrix theory[28].

Lemma 4.3: Let $x, y \in C([0,T], \mathbb{R}^n)$. If there exists some norm on $\mathbb{R}^n$ such that

$$\|\dot{x}(t)\| \leq \gamma \|\dot{y}(t)\| + l_1 \|x(t)\| + l_2 \|y(t)\| \qquad [4.2]$$

for some positive numbers $l_1, l_2 < \infty$ and $\gamma < 1$ then there exists a norm $\| \cdot \|_b$ on $C([0,T], \mathbb{R}^n)$ such that

$$\|\dot{x}\|_b \leq \alpha \|\dot{y}\|_b + l_1 \|x(0)\| + l_2 \|y(0)\| \qquad [4.3]$$

for some positive number $\alpha < 1$. ∎

Proof of Lemma 4.3:

Substituting $\int_0^t \dot{x}(\tau)d\tau + x(0)$ for $x(t)$ in Eqn. (4.2) and performing an analogous substitution for $y(t)$, multiplying the entire equation by e^{-bt}, and moving the norms inside the integral yields:

$$e^{-bt}\|\dot{x}(t)\| \le \gamma e^{-bt}\|\dot{y}(t)\| + l_1 e^{-bt}\int_0^t \|\dot{x}(\tau)\|d\tau + l_1 e^{-bt}\|x(0)\| \quad [4.4]$$

$$+ \; l_2 e^{-bt}\int_0^t \|\dot{y}(\tau)\|d\tau + l_2 e^{-bt}\|y(0)\|.$$

Let $\|\cdot\|_b$ be defined by $\|f\|_b \equiv \max_{[0,T]} e^{-bt}\|f(t)\|$. This is a norm on $C([0,T], \mathbb{R}^n)$ for any finite positive number $b > 0$ and is equivalent to the uniform norm on $C([0,T], \mathbb{R}^n)$. Then Eqn. (4.4) implies

$$\|\dot{x}\|_b \le \gamma\|\dot{y}\|_b + \max_{[0,T]}[\; l_1 e^{-bt}\int_0^t e^{b\tau}d\tau \; \|\dot{x}\|_b + l_1 e^{-bt}\|x(0)\| \; +$$

$$l_2 e^{-bt}\int_0^t e^{b\tau}d\tau \; \|\dot{y}\|_b + l_2 e^{-bt}\|y(0)\| \;].$$

And since $e^{-bt}\int_0^t e^{b\tau}d\tau \le \dfrac{1}{b}$, then for $b > l_1$ we can write

$$\|\dot{x}\|_b \le \frac{\gamma + l_2 b^{-1}}{1 - l_1 b^{-1}}\|\dot{y}\|_b + l_1\|x(0)\| + l_2\|y(0)\|. \quad [4.5]$$

In this case γ is less than 1, so there exists a finite B for which $\dfrac{(\gamma + l_2 B^{-1})}{1 - l_1 B^{-1}} = \alpha < 1$. Let the b in Eqn. (4.5) be set equal to this B to get

$$\|\dot{x}\|_B \le \alpha\|\dot{y}\|_B + l_1\|x(0)\| + l_2\|y(0)\|. \quad [4.6]$$

which completes the proof. ∎

Now we prove the following WR convergence theorem for systems of equations of the form of Eqn. (2.10).

<u>Theorem 4.1:</u>   If, in addition to the assumptions of Eqn. (2.10), $C(x(t),u(t)) \in \mathbb{R}^{n \times n}$ is strictly diagonally dominant uniformly over all $x(t) \in \mathbb{R}^n$

and $u(t) \in \mathbb{R}^r$ and Lipschitz continuous with respect to $x(t)$ for all $u(t)$, and $x^0(t)$ is differentiable, then the sequence of waveforms $\{x^k\}$ generated by the Gauss-Seidel or Gauss-Jacobi WR algorithm will converge uniformly to the solution of Eqn. (2.10) for all bounded intervals $[0,T]$.∎

Proof of Theorem 4.1:

We will present the proof only for the Gauss-Seidel WR algorithm, as the proof for the Gauss-Jacobi case is almost identical. Let x^{k+1} and x^k be the $(k + 1)^{st}$ and k^{th} iterates generated by applying the Gauss-Seidel WR algorithm to Eqn. (2.10). We can define the following from Eqn. (2.10):

$$\hat{C}_{ij}(x^{k+1}, x^k, u) = C_{ij}(x_1^{k+1}, ..., x_i^{k+1}, x_{i+1}^k, ..., x_n^k, u),$$

and

$$\hat{f}_i(x^{k+1}, x^k, u) = f_i(x_1^{k+1}, ..., x_i^{k+1}, x_{i+1}^k, ..., x_n^k, u).$$

Note that by Lemma 4.1, the matrix $\hat{C}$ is diagonally dominant because C in Eqn. (2.10) is diagonally dominant.

Then define L_{k+1}, D_{k+1}, and U_{k+1} such that $\hat{C}(x^{k+1}, x^k, u) = L_{k+1} + D_{k+1} - U_{k+1}$ and L_{k+1} is strictly lower triangular, U_{k+1} is upper triangular, and D_{k+1} is diagonal. The equations for one iteration of the Gauss-Seidel WR algorithm applied to Eqn. (2.10) can then be written in matrix form as

$$(L_{k+1} + D_{k+1})\dot{x}^{k+1} + U_{k+1}\dot{x}^k = \hat{f}(x^{k+1}, x^k, u)].$$

Rearranging the iteration equation yields

$$\dot{x}^{k+1} = (L_{k+1} + D_{k+1})^{-1}[U_{k+1}\dot{x}^k + \hat{f}(x^{k+1}, x^k, u)]. \qquad [4.7]$$

Taking the difference between Eqn. (4.7) at iteration $k + 1$ and at iteration $j + 1$ yields

$$\dot{x}^{k+1} - \dot{x}^{j+1} = (L_{k+1} + D_{k+1})^{-1}U_{k+1}x^k - (L_{j+1} + D_{j+1})^{-1}U_{j+1}x^j + \quad [4.8]$$

$$(L_{k+1} + D_{k+1})^{-1}\hat{f}(x^{k+1}, x^k, u) - (L_{j+1} + D_{j+1})^{-1}\hat{f}(x^{j+1}, x^j, u).$$

Using the Lipschitz continuity of $\hat{f}$ and the fact that $\|(L_{k+1} + D_{k+1})^{-1}\| < K$ for some $K < \infty$ independent of x and k (because $C(x,u)$ is uniformly diagonally dominant with respect to x and its inverse is bounded) in Eqn. (4.8) leads to

$$\|\dot{x}^{k+1}(t) - \dot{x}^{j+1}(t)\| \leq l_1 K \|x^{k+1}(t) - x^{j+1}(t)\| + l_2 K \|x^k(t) - x^j(t)\| + \quad [4.9]$$

$$\| (L_{k+1} + D_{k+1})^{-1} - (L_{j+1} + D_{j+1})^{-1} \| \, \|\hat{f}(x^{j+1}, x^j, u)\| +$$

$$\| (L_{k+1} + D_{k+1})^{-1}U_{k+1}\dot{x}^k(t) - (L_{j+1} + D_{j+1})^{-1}U_{j+1}\dot{x}^j(t) \|$$

where l_1 is the Lipschitz constant of $\hat{f}$ with respect to its first argument, and l_2 is the Lipschitz constant of $\hat{f}$ with respect to its second argument. That $C(x,u)$ is uniformly diagonally dominant and Lipschitz continuous with respect to x for all u implies $(L_k + D_k)^{-1}$ and $(L_k + D_k)^{-1}U_k$ are also Lipschitz continuous in the same manner. It then follows that there exist some positive finite numbers K_1, K_2, K_3, K_4 such that

$$\|\dot{x}^{k+1}(t) - \dot{x}^{j+1}(t)\| = l_1 K \|x^{k+1}(t) - x^{j+1}(t)\| + l_2 K \|x^k(t) - x^j(t)\| + \quad [4.10]$$

$$[K_3 \|x^{k+1}(t) - x^{j+1}(t)\| + K_4 \|x^k(t) - x^j(t)\|] \, \|\hat{f}(x^{j+1}, x^j, u)\| +$$

$$[K_1 \|x^{k+1}(t) - x^{j+1}(t)\| + K_2 \|x^k(t) - x^j(t)\|] \, \|\dot{x}^k(t)\| + \gamma \|\dot{x}^k(t) - \dot{x}^j(t)\|$$

where K_1 is the Lipschitz constant of $(L_k + D_k)^{-1}U_k$ with respect to its first x argument (see definition of L_k, U_k and D_k above), K_2 is the Lipschitz constant with respect to the second x argument, K_3 and K_4 are the Lipschitz constants for $(L_k + D_k)^{-1}$ with respect to its first and second x arguments, and γ is such that $\| (L_k + D_k)^{-1}U_k \| < \gamma < 1$ independent of k (by Lemma 4.2).

To establish a bound on the terms in Eqn. (4.10) involving $\| \dot{x}^k(t) \|$ and $\| \hat{f}(x^{j+1}, x^j, u) \|$, it is necessary to show that the $\dot{x}^k$'s and therefore the x^k's and $\hat{f}(\bullet)$'s are bounded a $priori$. We prove such a bound exists in the following lemma.

Lemma 4.4: If $C(x,u)$ in Eqn. (2.10) is strictly diagonally dominant and Lipschitz continuous then the $\dot{x}^k(t)$'s produced by Algorithm 4.1 are bounded independent of k.∎

Proof of Lemma 4.4

If $\| \bullet \|$ is the l_∞ norm on $\mathbb{R}^n$, by Lemma 4.2 $\| (L_{k+1} + D_{k+1})^{-1}U_{k+1} \| < 1$. From Eqn. (4.7),

$$\| \dot{x}^{k+1}(t) \| \leq \gamma \| \dot{x}^k(t) \| + \| (L_{k+1} + D_{k+1})^{-1} \| \, \| \hat{f}(x^{k+1}(t), x^k(t), u) \| \quad [4.11]$$

for some positive number $\gamma < 1$. As $f(x, u)$ is globally Lipschitz continuous with respect to x, there exist finite positive constants l_1, l_2 such that

$$\| \hat{f}(x, y, u) - \hat{f}(w, z, u) \| < l_1 \| x - w \| + l_2 \| y - z \| \quad [4.12]$$

for all $u, x, y, w, z \in \mathbb{R}^n$. From Eqn. (4.11) and Eqn. (4.12) and using the fact that $\| (L_{k+1} + D_{k+1})^{-1} \|$ is bounded by some $K < \infty$ for all k:

$$\| \dot{x}^{k+1}(t) \| \leq \gamma \| \dot{x}^k(t) \| + l_1 K \| x^{k+1}(t) \| + l_2 K \| x^k(t) \| + K \| \hat{f}(0,0,u) \|. \quad [4.13]$$

Eqn. (4.13) is of the form to apply a slightly modified Lemma 4.3. Therefore there exists some $\| \cdot \|_b$ such that

$$\| \dot{x}^{k+1} \|_b \leq \alpha \| \dot{x}^k \|_b + (l_1 K + l_2 K) \| x(0) \| + K \| \hat{f}(0, 0, u) \| \quad [4.14]$$

where $\alpha < 1$. This implies that

$$\| \dot{x}^{k+1} \|_b \leq \frac{1}{1-\alpha}[(l_1 K + l_2 K) \| x(0) \| + K \| \hat{f}(0, 0, u) \|] + (\alpha)^k \| \dot{x}^0 \|_b \quad [4.15]$$

for all k. Then, since $\| \dot{x}^0 \|_b$ is bounded by assumption, and $\| \dot{x}^{k+1} \|_b = \max_{[0,T]} e^{-bt} \| \dot{x}^{k+1}(t) \|$,

$$\| \dot{x}^{k+1}(t) \| \leq \quad [4.16]$$

$$e^{bT}[\frac{1}{1-\alpha}[(l_1 K + l_2 K) \| x(0) \| + K \| \hat{f}(0, 0, u) \|] + \| \dot{x}^0 \|_b] = \tilde{M}$$

which proves the lemma. ∎

We now continue with the proof of Theorem 4.1. In Lemma 4.4 it was proved that $\| \dot{x}^k(t) \|$ is bounded *a priori* by some $\tilde{M}$. This implies $x^k(t)$ is bounded on $[0,T]$. Using the Lipschitz-continuity property of $\hat{f}$, a bound, $\tilde{N}$, can be derived for $\| \hat{f}(x^{k+1}(t), x^k(t), u) \|$. Applying these bounds to Eqn. (4.10) we get

$$\| \dot{x}^{k+1}(t) - \dot{x}^{j+1}(t) \| \leq \gamma \| \dot{x}^k(t) - \dot{x}^j(t) \| + \quad [4.17]$$

$$(l_1 K + K_1 \tilde{M} + K_3 \tilde{N}) \| x^{k+1}(t) - x^{j+1}(t) \| + (l_2 K + \tilde{M} K_2 + K_4 \tilde{N}) \| x^k(t) - x^j(t) \|$$

where $\gamma < 1$. Eqn. (4.17) is of the form to apply Lemma 4.3. As $x^k(0) - x^j(0) = 0$ for all k, j, Lemma 4.3 implies

$$\| \dot{x}^{k+1} - \dot{x}^{j+1} \|_b \leq \alpha \| \dot{x}^k - \dot{x}^j \|_b \quad [4.18]$$

for some $0 < b < \infty$ and for some $\alpha < 1$. As $C([0,T], \mathbb{R}^n)$ is complete in any one of the b norms, by the contraction mapping theorem $\dot{x}^k$ converges to some $\dot{x} \in C([0,T], \mathbb{R}^n)$ which is a fixed point of Eqn. (4.7). Any fixed point $\dot{x}$ of Eqn. (4.7) is a solution to Eqn. (2.10) if $x(0) = x_0$. As $x^k(0) = x_0$ for all k, $\dot{x}^k$ converges to the unique solution of Eqn. (2.10). The sequence $\{x^k\}$ also converges, and this follows from the fact that integration from 0 to T, which maps $\dot{x}(t)$ to $x(t)$, is a bounded continuous function. ∎

SECTION 4.3 - WAVEFORM RELAXATION-NEWTON METHODS

The WR algorithm is an extension to function spaces of the relaxation methods used to solve linear and nonlinear systems. It is also possible to extend the Newton-Raphson algorithm, and its function-space extension also has practical applications. In particular, it is possible to approximately solve the WR iteration equations with one iteration of the waveform Newton algorithm, and this is the function-space extension of the relaxation-Newton methods described in Section 3.3. In this section we will derive the function-space Newton method as applied to systems of the form of Eqn. (2.10) and prove that the method has *global* convergence properties. We will then apply this method in conjunction with the WR algorithm to generate the waveform relaxation-Newton (WRN) algorithm.

In order to derive a function-space extension to the Newton-Raphson algorithm, let $F(x)$ (from Eqn. (2.10)) be defined by

$$F(x) = C(x, u)\dot{x} - f(x, u) = 0 \quad x(0) = x_0 \qquad [4.19]$$

where $x:[0,T] \rightarrow \mathbb{R}^n$; $u:[0,T] \rightarrow \mathbb{R}^r$ and is piecewise continuous; $C: \mathbb{R}^n x \mathbb{R}^r \rightarrow \mathbb{R}^{nxn}$ is such that $C(x, u)^{-1}$ exists and is uniformly bounded with respect to x, u; and $f: \mathbb{R}^n x \mathbb{R}^r \rightarrow \mathbb{R}^n$ is globally Lipschitz continuous with respect to x for all u. Applying the Newton-Raphson algorithm to find an x such that $F(x) = 0$ given some initial guess x^0 we get

$$x^{k+1} = x^k - J_F^{-1}(x^k)F(x^k) \tag{4.20}$$

where $J_F(x)$ is the Frechet derivative of $F(x)$ with respect to x. Note that in this case $J_F(x)$ is a matrix-valued function on $[0,T]$. That is, $J_F(x)$ is a matrix of waveforms.

Using the definition of the Frechet derivative, we can compute $J_F(x)$,

$$\lim_{h \to 0} (1/\|h\|) \| F(x + h) - F(x) - J_F(x)(h) \| = 0. \tag{4.21}$$

Evaluating this limit for the $F(x)$ given in Eqn. (4.19) we get

$$F(x + h) - F(x) = \tag{4.22}$$

$$C(x + h, u)(\dot{x} + \dot{h}) - C(x, u)\dot{x} - f(x + h, u) + f(x, u),$$

and approximating to order $\|h\|^2$

$$F(x + h) - F(x) = C(x,u)\dot{h} + \frac{\partial C(x,u)}{\partial x}h\dot{x} - \frac{\partial f(x,u)}{\partial x}h + O(\|h\|^2). \tag{4.23}$$

As Eqn. (4.21) applies only in the limit as $h \to 0$, Eqn. (4.23) implies

$$J_F(x)h = C(x,u)\dot{h} + \frac{\partial C(x,u)}{\partial x}h\dot{x} - \frac{\partial f(x,u)}{\partial x}h. \tag{4.24}$$

Substituting the computed derivative into Eqn. (4.20) and rearranging we get

$$C(x^k,u)\dot{x}^{k+1} + \frac{\partial C(x^k, u)}{\partial x}(x^{k+1} - x^k)\dot{x}^k = \tag{4.25}$$

$$f(x^k,u) + \frac{\partial f(x^k,u)}{\partial x}(x^{k+1} - x^k),$$

$$x^{k+1}(0) = x_0.$$

We will refer to Eqn. (4.25) as the waveform Newton(WN) algorithm for solving Eqn. (2.10). It is, however, just the function-space extension of the classical Newton-Raphson algorithm.

Newton algorithms converge quadratically when the iterated value is close to the correct solution, but they do not, in general, have global convergence properties. The WN algorithm, along with inheriting the locally quadratic convergence properties of general Newton methods, will also converge globally, given the mild assumptions on the behavior of $\dfrac{\partial C(x,u)}{\partial x}$ stated in the following theorem:

Theorem 4.2: For any system of the form of Eqn. (2.10) in which $\dfrac{\partial C(x,u)}{\partial x}$ is Lipschitz continuous with respect to x for all u and f is continuously differentiable, the sequence $\{x^k\}$ generated by the WN algorithm converges uniformly to the solution of Eqn. (2.10). ∎

Proof of Theorem 4.2

For this proof of the convergence of the waveform Newton method we will assume that $C(x,u)$ is independent of x and u, as the proof for the general case is much more involved, and does not provide much further insight into the nature of the convergence. For the case $C(x,u) = C$, Eqn. (4.25) can be simplified to

$$\dot{x}^{k+1} = C^{-1}f(x^k,u) + C^{-1}\frac{\partial f(x^k,u)}{\partial x}(x^{k+1} - x^k). \qquad [4.26]$$

Taking the difference between Eqn. (4.26) at iteration $k + 1$ and the exact solution and substituting $(x^{k+1} - x) + (x - x^k)$ for $x^{k+1} - x^k$ yields

$$\dot{x}^{k+1} - \dot{x} \qquad\qquad\qquad\qquad\qquad\qquad\qquad\qquad\qquad [4.27]$$

$$= C^{-1}[f(x^k,u) - f(x,u)] + C^{-1}[\frac{\partial f(x^k, u)}{\partial x}((x^{k+1} - x) + (x - x^k))].$$

The matrix C has a bounded inverse (by the assumptions following Eqn. (2.10)) and by assumption f is continuously differentiable on $[0,T]$ and is therefore

Lipschitz continuous. From these facts it follows that $C^{-1}\dfrac{\partial f(x,u)}{\partial x}$ is bounded by some constant l_1. With this bound,

$$\|\dot{x}^{k+1} - \dot{x}\| < l_1 \|x^k - x\| + l_1 \|x^{k+1} - x\| + l_1 \|x^k - x\|. \qquad [4.28]$$

Lemma 4.3 can be applied to Eqn. (4.28), where in this case the γ of Lemma 4.3 is zero. Therefore there exists some $b < \infty$ and $\alpha < 1$ such that

$$\|\dot{x}^{k+1} - \dot{x}\|_b \leq \alpha \|\dot{x}^k - \dot{x}\|_b. \qquad [4.29]$$

From this inequality it follows that the sequence $\{\dot{x}^k\}$ converges to $\dot{x}$, the fixed point of Eqn. (4.26). Given $x^k(0) = x_0$ for all k, $\{x^k\}$ converges to the solution of Eqn. (2.10) on any bounded interval. ∎

As mentioned in the introduction, it is possible to combine the waveform Newton method derived above with the WR algorithm to construct the waveform extension of the relaxation-Newton algorithms presented in Section 3.3[19]. The WR iteration equations are solved approximately by performing one step of this Newton method with each waveform relaxation iteration, to yield Algorithm 4.2, the waveform relaxation-Newton algorithm (WRN). Like Algorithm 4.1, each equation in Algorithm 4.2 has only one unknown variable x_i^k, but in this case each of the nonlinear equations has been replaced by a simpler time-varying linear equation.

Given the global convergence properties of both the original WR and the WN algorithms, it is not surprising that the WRN algorithm has global convergence properties. We will state the convergence theorem, but will not present the proof because it quite similar to the proof of the basic WR and WN convergence theorems.

Theorem 4.3: If, in addition to the assumptions of Theorem 4.1, $\dfrac{\partial C(x,u)}{\partial x}$ is Lipschitz continuous with respect to x for all u and f is continuously differentiable, then the sequence $\{x^k\}$ generated by the Gauss-Seidel or Gauss-

Algorithm 4.2 - (WRN Gauss-Seidel Algorithm for solving Eqn. (2.10))
The superscript k denotes the iteration count, the subscript $i \in \{1, ..., N\}$ denotes the component index of a vector and ϵ is a small positive number.
$k \leftarrow 0$;
guess waveform $x^0(t)$; $t \in [0,T]$ such that $x^0(0) = x_0$
   (for example, set $x^0(t) = x_0$, $t \in [0,T]$);
repeat {
  $k \leftarrow k + 1$
  for all (i in N) {
   solve

$$\sum_{j=1}^{i-1} C_{ij}(x_1^k, ..., x_{i-1}^k, x_i^{k-1}, ..., x_n^{k-1}, u)\dot{x}_j^k +$$

$$\frac{\partial C_{ii}(x_1^k, ..., x_{i-1}^k, x_i^{k-1}, ..., x_n^{k-1}, u)}{\partial x_i}(x_i^k - x_i^{k-1})\dot{x}_i^k +$$

$$\sum_{j=i+1}^{n} C_{ij}(x_1^k, ..., x_{i-1}^k, x_i^{k-1}, ..., x_n^{k-1}, u)\dot{x}_j^{k-1} -$$

$$f_i(x_1^k, ..., x_{i-1}^k, x_i^{k-1}, ..., x_n^{k-1}, u) -$$

$$\frac{\partial f_i(x_1^k, ..., x_{i-1}^k, x_i^{k-1}, ..., x_n^{k-1}, u)}{\partial x_i}(x_i^k - x_i^{k-1}) = 0$$

   for ($x_i^k(t)$; $t \in [0,T]$), with the initial condition $x_i^k(0) = x_{i_0}$.
  }
 } until ($\| x^k - x^k \| \leq \epsilon$)
∎ .

Jacobi WRN algorithm converges to the solution of Eqn. (2.10) on all bounded intervals $[0,T]$.∎

The linear time-varying systems generated by the WRN algorithm are easier to solve numerically than the nonlinear iteration equations of the basic WR algorithm. For example, if an implicit multistep integration method is used to solve such a system, the implicit algebraic equations that the multistep method generates will be linear[20][66]. In addition, linear time-varying systems can be solved with a variety of efficient numerical techniques other than the standard discretization methods, such as collocation[58] and spectral methods[22].

SECTION 4.4 - NONSTATIONARY WR ALGORITHMS

Algorithm 4.1 is stationary in the sense that the equations which define the iteration process do not change with the iterations. A straightforward generalization is to allow these iteration equations to change, and to consider under what conditions the relaxation still converges [13]. There are two major reasons for studying nonstationary algorithms. The solution of the ordinary differential equations in the inner loop of Algorithm 4.1 cannot be obtained exactly. Instead numerical methods compute the solution with some error which is in general controlled, but which cannot be eliminated. However, the discrete approximation can be interpreted as the exact solution to a perturbed system. Since the approximation changes with the solutions, the perturbed system changes with each iteration. Hence, practical implementations of WR that must compute the solution to the iteration equations approximately can be interpreted as nonstationary methods.

The second reason for studying nonstationary methods is that they can be used to improve the computational efficiency of the basic WR algorithm. One approach would be to improve the accuracy of the computation of the iteration equations as the relaxation approaches convergence. In this way, accurate solutions to the original system would still be obtained, but unnecessarily accurate computation of the early iteration waveforms, which are usually far from the final solution, is avoided.

In this section we show that nonstationary WR algorithms converge as a direct consequence of the contraction-mapping property of the original WR algorithm. That is, given mild assumptions about the relationship between a general stationary contraction map and a nonstationary map, the nonstationary map will produce a sequence that will converge to within some tolerance. And, if in the limit as $k \to \infty$ the nonstationary map approaches the stationary map, then the sequence generated by the nonstationary map will converge to the fixed point of the original map. In later sections we will lean on these results to guarantee the convergence of implementations of WR-based algorithms.

Theorem 4.4: Let Y be a Banach space and F, $F^k:Y \to Y$ be such that there exists a $z \in Y$, is such that $z = F(z)$. Define $y^k, \tilde{y}^k \in Y$ such that $y^{k+1} = F(y^k)$ and $\tilde{y}^{k+1} = F^k(\tilde{y}^k)$. If F is a contraction mapping with contraction factor γ (See section 4.2) and there exists a sequence of positive numbers $\{\delta^k\}$ such that $\| F(y) - F^k(y) \| \le \delta^k$ for all $y \in Y$, then for any $\varepsilon > 0$ there exists a $\delta < 1$ such that if $\delta^k < \delta$ for all k then $\lim_{k\to\infty} \| \tilde{y}^k - \tilde{y}^{k-1} \| < \varepsilon$ and $\lim_{k\to\infty} \| z - \tilde{y}^k \| < \dfrac{\delta}{1-\gamma}$. Futhermore, if $\lim_{k\to\infty} \delta^k \to 0$ then $\lim_{k\to\infty} \| z - \tilde{y}^k \| = 0.$ ∎

Proof of Theorem 4.4

Taking the norm of the difference between the kth and $(k + 1)^{st}$ iteration of the nonstationary algorithm we get:

$$\| \tilde{y}^{k+1} - \tilde{y}^k \| < \| F^k(\tilde{y}^k) - F^{k-1}(\tilde{y}^{k-1}) \|. \qquad [4.30]$$

Given that $\| F^k(y) - F(y) \| < \delta^k$ for all $y \in Y$,

$$\| \tilde{y}^{k+1} - \tilde{y}^k \| \le \| F(\tilde{y}^k) - F(\tilde{y}^{k-1}) \| + \delta^k + \delta^{k-1}. \qquad [4.31]$$

Using the contraction property of F,

$$\| \tilde{y}^{k+1} - \tilde{y}^k \| < \gamma \| \tilde{y}^k - \tilde{y}^{k-1} \| + \delta^k + \delta^{k-1}. \qquad [4.32]$$

Unfolding the iteration equation,

$$\| \tilde{y}^{k+1} - \tilde{y}^k \|_b \le \delta^{k-1} + \delta^k + \sum_{i=1}^{k} \gamma^{k-i}(\delta^i + \delta^{i-1}). \qquad [4.33]$$

If $\delta^k < \delta$ for all k, then from Eqn. (4.33)

$$\lim_{k\to\infty} \|\tilde{y}^{k+1} - \tilde{y}^{k}\| \leq 2\delta(1 + \frac{1}{1-\gamma}).$$ [4.34]

As $\gamma < 1$, $\lim_{k\to\infty} \|\tilde{y}^{k+1} - \tilde{y}^{k}\|$ can be made as small as desired by reducing δ, which proves the first part of Theorem 4.4.

Let z be the fixed point of F. The difference between the computed and the exact solution at the $(k + 1)^{st}$ iteration is

$$\|\tilde{y}^{k+1} - z\| = \|F^k(\tilde{y}^k) - F(z)\|.$$ [4.35]

Again using the contractive property of F and the fact that $\|F(y) - F^k(y)\| \leq \delta^k$,

$$\|\tilde{y}^{k+1} - z\| = \gamma\|\tilde{y}^k - z\| + \delta^k.$$ [4.36]

Summing and taking the limit,

$$\lim_{k\to\infty} \|\tilde{y}^{k+1} - z\| \leq \frac{\delta}{1-\gamma},$$ [4.37]

which completes the proof of the first statement of Theorem 4.4. The second statement of the theorem follows from almost identical arguments. ∎

In Section 4.2 we proved the WR iteration was a contraction mapping in the appropriate norm $\| \cdot \|_b$ on $C([0,T], \mathbb{R}^n)$ where B depended on the problem. To repeat the result from that section, it was shown that

$$\|\dot{x}^{k+1} - \dot{x}^{j+1}\|_b \leq \alpha\|\dot{x}^k - \dot{x}^j\|_b$$

where $\alpha < 1$. This WR convergence result and Theorem 4.4 imply that any "reasonable" approximation method used to solve the WR iteration equations will produce a converging sequence of waveforms *provided* the errors in the approximation are driven to zero. In addition, Theorem 4.4 indicates that it will be difficult to determine *a priori* how accurately the iteration equations must be solved in order to guarantee convergence to within a given tolerance, because an estimate of the contraction factor of the WR algorithm is required.

From Theorem 4.1, the WR is a contraction mapping with respect to $\dot{x}(t)$ in a B norm. Theorem 4.4 then implies that the WR iteration equations must be solved accurately with respect to $\dot{x}(t)$ in this B norm if the iterations are to converge. There is a more cumbersome proof of the WR convergence theorem[12] in which it is shown that the WR algorithm is a contraction in $x(t)$. The proof uses a larger B norm than the one used in the proof of Theorem 4.1, and the size of this B is a function of the magnitude of the off-diagonal terms of $C(x,u)$. With such a result, Theorem 4.4 implies that it is only necessary to control errors in the computation of $x(t)$ to guarantee iteration convergence. However, convergence in a larger B norm is in some sense a weaker type of convergence. So, in the case where $C(x,u)$ has nonzero off-diagonal terms, it is expected that more rapid convergence would be achieved if the $x^k(t)$'s are computed in a way that also guarantees that the $\dot{x}^k(t)$'s are globally accurate.

CHAPTER 5 - ACCELERATING WR CONVERGENCE

In Chapter 7, several techniques used to improve the efficiency and robustness of the WR algorithm when it is applied to simulating MOS circuits will be described. In this chapter the theoretical background for three of these techniques will be presented. We will first analyze nonuniformity in WR convergence, which explains why breaking the simulation interval into pieces, called windows, can be used to reduce the number of relaxation iterations required to achieve convergence. Then we will consider how to partition large systems into subsystems in such a way that the WR algorithm will converge rapidly. Finally, we will briefly address a well-known issue for Gauss-Seidel relaxation algorithms, that of ordering the equations.

SECTION 5.1 - UNIFORMITY OF WR CONVERGENCE

The convergence theorem presented in Section 4.2 guarantees that the WR algorithm is a contraction mapping in an exponentially weighted norm. In this section, we will examine the implications of this choice of norm. First, the WR algorithm will be applied to two sample problems to demonstrate the different manners in which the algorithm converges. We will then prove that for a special class of systems WR converges in a uniform manner or, formally, that WR is a contraction in an unweighted norm for any finite time interval. Because most circuit problems do not generate systems in this special class, we will prove that the WR algorithm is a contraction in an unweighted norm for any system for which it converges, if the time interval is made short enough. This suggests that the number of iterations required to achieve WR convergence can be reduced by breaking the simulation interval into short pieces, and in Chapter 7 we will present an adaptive algorithm that attempts to exploit this property of WR.

Consider the following nonlinear ordinary differential equation in $x_1(t), x_2(t) \in \mathbb{R}$ with input $u \in \mathbb{R}$ that approximately describes the cross-coupled *nor* logic gate in Fig. 5.1a (the approximate equations represent a normalization that converts the simulation interval $[0,T]$ to $[0,1]$):

$$\dot{x}_1 = (5 - x_1) - x_1(x_2)^2 - 5x_1u \qquad\qquad [5.1]$$

$$\dot{x}_2 = (5 - x_2) - x_2(x_1)^2$$

$$x_1(0) = 5.0 \qquad x_2(0) = 0.$$

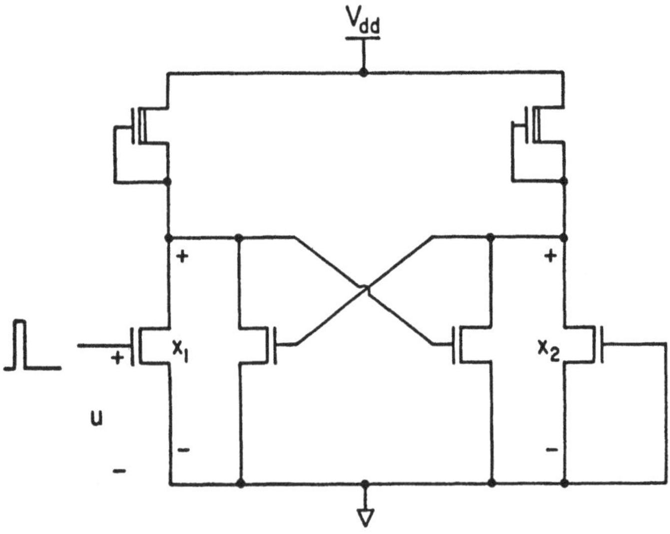

Figure 5.1a - Cross-Coupled Nor Gate

The Gauss-Seidel WR Algorithm given in Section 4.2 was used to solve for the behavior of the cross-coupled *nor* gate circuit approximated by the above small system of equations. Fig. 5.1b shows the input $u(t)$, the exact solution for

$x_1(t)$, and the relaxation iteration waveforms for $x_1(t)$ for the 5th, 10th and 20th iterations. The plots demonstrate a property typical of the WR algorithm when applied to systems with strong coupling: the difference between the iteration waveforms and correct solution is not reduced at every time point in the waveform. Instead, each iteration lengthens the interval of time, starting from zero, for which the waveform is close to the exact solution.

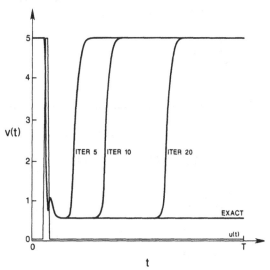

Figure 5.1b - WR Iteration from Cross-Coupled Nor Gate

As an example of "better" convergence, consider the following differential equation in x_1, x_2, x_3 with input u that approximately describes the shift register in Fig. 5.2a (here the simulation interval $[0,T]$ has been normalized to $[0,1]$):

$$\dot{x}_1 = (5.0 - x_1) - x_1(u)^2 - (x_1 - x_2) \qquad [5.2]$$

$$\dot{x}_2 = (x_1 - x_2)$$

$$\dot{x}_3 = (5.0 - x_3) - x_3(x_2)^2$$

$$x(0) = 0.$$

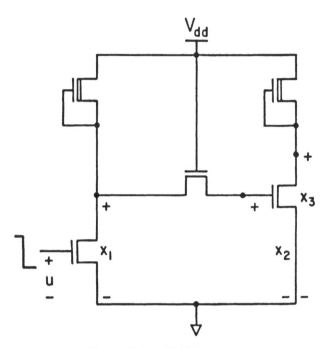

Figure 5.2a - Shift Register

The Gauss-Seidel WR Algorithm given in Section 4.1 was used to solve the original system approximated by the above system of equations. The input $u(t)$, the exact solution for $x_1(t)$, and the waveforms for $x_1(t)$ computed from the first, second, and third iterations of the WR algorithm are plotted in Fig. 5.2b. In this example the difference between the iteration waveforms and the correct solution is reduced throughout the entire waveform.

Perhaps surprisingly, the behavior of the first example is consistent with the WR convergence theorem, even though that theorem states that the iterations converge uniformly. This is because it was proved that the WR method is a contraction map in the following nonuniform norm on $C([0,T], \mathbb{R}^n)$:

$$\max_{[0,T]} e^{-bt} \| f(t) \|$$

where $b > 0$, $f(t) \in \mathbb{R}^n$, and $\| \cdot \|$ is a norm on $\mathbb{R}^n$. Note that $\| f(t) \|$ can increase as e^{bt} without increasing the value of this function-space norm. If $f(t)$ grows slowly, or is bounded, it is possible to reduce the function-space norm by

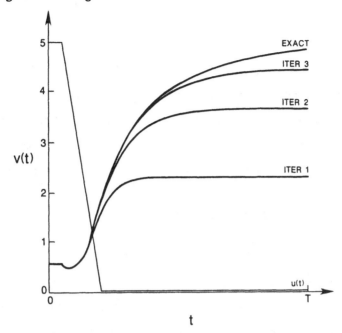

Figure 5.2b – WR Iteration from Shift Register

reducing $\|f(t)\|$ only on some small interval in $[0,T]$, though it will be necessary to increase this interval to further decrease the norm. The waveforms in the more slowly converging example above converge in just this way; the function-space norm is decreased after every iteration of the WR algorithm because significant errors are reduced over larger and larger intervals of time.

The examples above lead to the following definition:

Definition 5.1: A differential system of the form given in Eqn. (2.10) is said to have the *strict WR contractivity property* on $[0,T]$, if the WR algorithm applied to the system contracts in a uniform norm on $[0,T]$, i.e.

$$\max_{[0,T]} \|x^{k+1}(t) - x^k(t)\| \; < \; \max_{[0,T]} \|x^k(t) - x^{k-1}(t)\| \qquad [5.3]$$

where $x^k(t) \in \mathbb{R}^n$ on $t \in [0,T]$ is the kth iterate of Algorithm 4.1 and $\| \cdot \|$ is any norm on $\mathbb{R}^n$. If the WR algorithm applied to the system is a contraction in a uniform norm on $[0,T]$ for any $T > 0$ then the system has the strict WR contractivity property on $[0, \infty)$ ∎

For a system of equations to have the strict WR contractivity property on $[0, \infty)$ it must be more than just loosely coupled. In addition, the decomposed equations solved at each iteration of the waveform relaxation must be well-damped, so that errors due to the decomposition die off in time, instead of accumulating or growing. As an example, we will prove that a system in normal form,

$$\dot{x}(t) = f(x(t), u(t)) \qquad x(0) = x_0 \qquad [5.4]$$

where $x(t) \in \mathbb{R}^n$ on $t \in [0,T]$; $u(t) \in \mathbb{R}^r$ on $t \in [0,T]$ is piecewise continuous; and $f: \mathbb{R}^n x \mathbb{R}^r \rightarrow \mathbb{R}^n$ is globally Lipschitz continuous; will have the strict WR contractivity property on $[0,T]$ for any $T < \infty$ if f has a property we refer to as diagonally dominant negative monotonicity. This property, which we precisely define below, just implies that the original system is loosely coupled and the decomposed equations generated by a WR algorithm are well-damped (A similar result in a different setting can be found in [61]).

Definition 5.2: Let $f(x, u)$ be a continuous map from $\mathbb{R}^n x \mathbb{R}^r \rightarrow \mathbb{R}^n$ where $x \in \mathbb{R}^n$, $u \in \mathbb{R}^r$; and f is globally Lipschitz continuous with respect to x for all $u \in \mathbb{R}^r$. Then, f is said to be negative monotone if there exists a positive number λ such that

$$(x - y) \bullet (f(x, u) - f(y, u)) \leq -\lambda(x - y) \bullet (x - y) \qquad [5.5]$$

(here $\bullet$ indicates a scalar product) for all $x, y \in \mathbb{R}^n$ and $u \in \mathbb{R}^r$. Let $v^i \in \mathbb{R}^n$ be the ith unit vector. Then f is said to be diagonally negative monotone if there exists a collection of positive numbers λ_i such that for all $i \in \{1,...,n\}$,

$$\varepsilon v^i \bullet (f(x + \varepsilon v^i, u) - f(x,u)) \leq -\lambda_i \varepsilon^2 \qquad [5.6]$$

for any positive $\varepsilon \in \mathbb{R}$, $x \in \mathbb{R}^n$ and $u \in \mathbb{R}^r$. If f is globally Lipschitz continous, there exist positive numbers l_{ij}, $i,j \in [1, ..., n]$ such that for any $\varepsilon \in \mathbb{R}$

$$\| v^i \cdot (f(x + \epsilon v^j, u) - f(x,u)) \| \leq l_{ij} |\epsilon| \qquad [5.7]$$

for all $x \in \mathbb{R}^n$, $u \in \mathbb{R}^r$. A mapping f, is a diagonally dominant negative monotone if f is a strictly diagonally negative monotone and $\lambda_i > \sum_{j \neq i} l_{ij}$ where λ_i is as in Eqn. (5.6). (This is a stricter definition than in previous literature[30]) ■.

To prove the theorem about diagonally dominant negative monotone systems we will use the following lemma.

Lemma 5.1: Let k, $x(t)$, $\dot{x}(t) \in \mathbb{R}$ be such that

$$x(t)\dot{x}(t) \leq -\lambda x(t)x(t) + kx(t) \qquad x(0) = 0 \qquad [5.8]$$

for some positive number λ. Then $|x(t)| < |k|\lambda^{-1}$ for all $t > 0$.■

Proof of Lemma 5.1:

Substituting $\dfrac{d}{dt}|x(t)|^2$ for $2x(t)\dot{x}(t)$ in Eqn. (5.8) and taking absolute values

$$\frac{d}{dt}|x(t)|^2 \leq -2\lambda |x(t)|^2 + 2|k| |x(t)|.$$

Therefore, $\dfrac{d}{dt}|x(t)| \leq -\lambda |x(t)| + |k|$ or $|x(t)| = 0$. This implies, by a theorem in differential inequalities[39], that

$$|x(t)| \leq \frac{|k|}{\lambda}(1 - e^{-\lambda t}) \leq \frac{|k|}{\lambda}$$

which proves the lemma. ■

We now prove the theorem.

Theorem 5.1: Let a system of equations of the form of Eqn. (5.4) be such that $f(x,u)$ is diagonally dominant negative monotone. Then the system has the strict WR contractivity property on $[0, T]$ for all $T < \infty$.■

Proof of Theorem 5.1:

Again we will only present the proof for the Gauss-Seidel case, but the result holds for the Gauss-Jacobi case also. The iteration equations for the Gauss-
Seidel WR algorithm applied to Eqn. (5.4) are, for each $i \in [1, ..., n]$,

$$\dot{x}_i^{k+1} = f_i(x_1^{k+1}, ..., x_i^{k+1}, x_{i+1}^k, ..., x_n^k, u) \qquad [5.9]$$

where x, u, and f are functions of time, but the dependence on time has been
dropped for notational convenience. Taking the difference between the kth and
$(k + 1)^{st}$ iteration for each $i \in [1, ..., n]$ yields

$$\dot{x}_i^{k+1} - \dot{x}_i^k = f_i(x_1^{k+1}, ..., x_i^{k+1}, x_{i+1}^k, ..., x_n^k, u) - f_i(x_1^k, ..., x_i^k, x_{i+1}^{k-1}, ..., x_n^{k-1}, u).$$

Multiplying through by $x_i^{k+1} - x_i^k$ and using the Lipschitz and diagonal negative
monotone properties of f we get

$$(x_i^{k+1} - x_i^k)(\dot{x}_i^{k+1} - \dot{x}_i^k) \leq -\lambda_i(x_i^{k+1} - x_i^k)(x_i^{k+1} - x_i^k) + \qquad [5.10]$$

$$\sum_{j=1}^{i-1} l_{ij}|x_j^{k+1} - x_j^k| \cdot |x_i^{k+1} - x_i^k| + \sum_{j=i+1}^{n} l_{ij}|(x_j^k - x_j^{k-1})| \cdot |x_i^{k+1} - x_i^k|$$

where l_{ij} and λ_i are as in Definition 5.2. From the estimate in Lemma 5.1,

$$|x_i^{k+1} - x_i^k| < \sum_{j=1}^{i-1} l_{ij}\lambda_i^{-1}|x_j^{k+1} - x_j^k| + \sum_{j=i+1}^{n} l_{ij}\lambda_i^{-1}|x_j^k - x_j^{k-1}|. \quad [5.11]$$

Let $A \in \mathbb{R}^{n \times n}$ be a matrix defined by $A_{ij} = l_{ij}\lambda_i^{-1}$ for all $i \neq j$ and $A_{ii} = 0$. Then
$A = L + U$ where L is strictly lower triangular, U is strictly upper triangular.
Rewriting Eqn. (5.11) in matrix form

$$(I - L)|x^{k+1} - x^k| \leq U|x^k - x^{k-1}| \qquad [5.12]$$

where $|x|$ is the vector whose elements are the absolute value of the elements of x, and the inequality holds for each element-by-element comparison. To show that Eqn. (5.12) implies $\|x^{k+1} - x^k\|_\infty < \|x^k - x^{k-1}\|_\infty$ requires a slightly complicated argument, as the inequality will not still hold if both sides of Eqn. (5.12) are multiplied by $(I - L)^{-1}$. Since $(I - L)$ is diagonally dominant with unit diagonal entries and negative lower-triangular off-diagonal entries, if r is a solution to $(I - L)|r| = U|x^k - x^{k-1}|$ then $|r| \geq |x^{k+1} - x^k|$. Given that f is diagonally dominant, $\|(I - L)^{-1}U\|_\infty < 1$ (Lemma 4.2), from which it follows that $|x^{k+1} - x^k| \leq r < |x^k - x^{k-1}|$. Then from Eqn. (5.12) we get

$$\max_{[0,T]} \|x^{k+1}(t) - x^k(t)\|_\infty < \max_{[0,T]} \|x^k(t) - x^{k-1}(t)\| \qquad [5.13]$$

for any $T < \infty$, which proves the theorem. ∎

As the crossed *nand* gate example indicates, many systems of interest do not have the strict WR contractivity property on $[0,T]$ for all $T < \infty$. However, we will prove that any system that satisfies the WR convergence theorem will also have the strict WR contractivity property on some nonzero interval.

Theorem 5.2: For any system of the form of Eqn. (2.10) which satisfies the assumptions of the WR convergence theorem (Theorem 4.1) there exists a $T > 0$ such that the system has the strict WR contractivity property on $[0,T]$.

Proof of Theorem 5.2

We prove the theorem only for the Gauss-Seidel WR algorithm but, as before, the theorem also holds for the Gauss-Jacobi case. Starting with Eqn. (4.8) and substituting x^k for x^{j+1},

$$\dot{x}^{k+1}(t) - \dot{x}^k(t) = \qquad [5.14]$$

$$(L_{k+1}(t) + D_{k+1}(t))^{-1}U_{k+1}(t)\dot{x}^k(t) - (L_k(t) + D_k(t))^{-1}U_k(t)\dot{x}^{k-1}(t) +$$

$$(L_{k+1}(t) + D_{k+1}(t))^{-1}\hat{f}(x^{k+1}, x^k, u) - (L_k(t) + D_k(t))^{-1}\hat{f}(x^k, x^{k-1}, u).$$

To simplify the notation, let $A_k(t)$, $B_k(t)$ ϵ $\mathbb{R}^{n \times n}$ be defined by $A_k(t) = (L_k(t) + D_k(t))^{-1}U_k(t)$, $B_k(t) = (L_k(t) + D_k(t))^{-1}$. It is important to keep in mind that $(L_k(t) + D_k(t))^{-1}U_k(t)$, and $(L_k(t) + D_k(t))^{-1}$ are functions of x^k and, by definition, so are $A_k(t)$ and $B_k(t)$. Expanding the above equation and integrating,

$$\int_0^t (\dot{x}^{k+1}(\tau) - \dot{x}^k(\tau))d\tau = \int_0^t A_{k+1}(\tau)(\dot{x}^k(\tau) - \dot{x}^{k-1}(\tau))d\tau + \qquad [5.15]$$

$$\int_0^t [A_{k+1}(\tau) - A_k(\tau)] \dot{x}^{k-1}(\tau) \, d\tau +$$

$$\int_0^t B_{k+1}(\tau)[\hat{f}(x^{k+1}(\tau), x^k(\tau), u(\tau)) - \hat{f}(x^k(\tau), x^{k-1}(\tau), u(\tau))]d\tau +$$

$$\int_0^t [B_{k+1}(\tau) - B_k(\tau)] \hat{f}(x^k(\tau), x^{k-1}(\tau), u(\tau))d\tau.$$

Integrating by parts and using the fact that $x^k(0) - x^{k-1}(0) = 0$,

$$x^{k+1}(t) - x^k(t) = A_{k+1}(t) [x^k(t) - x^{k-1}(t)] - \qquad [5.16]$$

$$\int_0^t \frac{d}{d\tau} A_{k+1}(\tau)[x^k(\tau) - x^{k-1}(\tau)]d\tau + \int_0^t [A_{k+1}(\tau) - A_k(\tau)] \dot{x}^{k-1} \, d\tau +$$

$$\int_0^t B_{k+1}(\tau)[\hat{f}(x^{k+1}(\tau), x^k(\tau), u(\tau)) - \hat{f}(x^k(\tau), x^{k-1}(\tau), u(\tau))]d\tau +$$

$$\int_0^t [B_{k+1}(\tau) - B_k(\tau)] \hat{f}(x^k(\tau), x^{k-1}(\tau), u(\tau))d\tau.$$

Taking norms, and using the Lipschitz continuity of f, $A_k(t)$, and $B_k(t)$, and the uniform boundedness of $B_k(t)$ in x (see Theorem 4.1):

$$\| x^{k+1}(t) - x^k(t) \| \quad - \int_0^t (l_1 K + K_1 \tilde{M} + K_3 \tilde{N}) \| x^{k+1}(\tau) - x^k(\tau) \| d\tau \quad \leq \quad [5.17]$$

$$\gamma \| x^k(t) - x^{k-1}(t) \| + \int_0^t (l_2 K + K_1 \tilde{M} + 2K_2 \tilde{M} + K_4 \tilde{N}) \| x^k(\tau) - x^{k-1}(\tau) \| d\tau$$

where l_1 , l_2 are the Lipschitz constants of $\hat{f}$ with respect to x^{k+1} and x^k respectively; K_1, K_2, K_3, K_4 are the Lipschitz constants for $A_{k+1}(t)$, $B_{k+1}(t)$ with respect to their x^{k+1} and x^k arguments respectively; $\gamma = \max_{x,k} \| (L_k + D_k)^{-1} U_k \| < 1$; and $\tilde{M}$, $\tilde{N}$ and K are the *a priori* bounds found in the proof of Theorem 4.1 on $\dot{x}^k$, $\hat{f}$, and $B_{k+1}(t)$ respectively. Note that $\frac{d}{d\tau} A_{k+1}(\tau) = \frac{d}{dx^{k+1}} A_{k+1}(\tau) \dot{x}^{k+1} + \frac{d}{dx^k} A_{k+1}(\tau) \dot{x}^k < K_1 \tilde{M} + K_2 \tilde{M}$. Moving the max (over t) norms outside the integrals and integrating yields

$$\max_{[0,T]} \| x^{k+1}(t) - x^k(t) \| \quad \leq \quad\quad\quad\quad [5.18]$$

$$\frac{\gamma + T(l_2 K + K_1 \tilde{M} + 2K_2 \tilde{M} + K_4 \tilde{N})}{1 - T(l_1 K + K_1 \tilde{M} + K_3 \tilde{N})} \max_{[0,T]} \| (x^k(t) - x^{k-1}(t)) \| .$$

Since $\gamma < 1$, a $T' > 0$ exists such that

$$\frac{\gamma + T'(l_2 K + K_1 \tilde{M} + 2K_2 \tilde{M} + K_4 \tilde{N})}{1 - T'(l_1 K + K_1 \tilde{M} + K_3 \tilde{N})} = \alpha < 1.$$

With this T', Eqn. (5.17) becomes

$$\max_{[0, T']} \| x^{k+1} - x^k \| \leq \alpha \max_{[0, T']} \| x^k - x^{k-1} \| \quad\quad [5.19]$$

for $\alpha < 1$, which proves the theorem. ∎

Theorem 5.2 guarantees that the WR algorithm is a contraction mapping in a uniform norm for any system, provided the interval of time over which the waveforms are computed is made small enough. This suggests that the interval of simulation $[0,T]$ should be broken up into *windows*, $[0, T_1], [T_1, T_2], ..., [T_{n-1}, T_n]$, where the size of each window is small enough so that the WR algorithm contracts uniformly throughout the entire window. The smaller the window is made, the faster the convergence. However, as the window size becomes smaller, the advantages of the waveform relaxation are lost. Scheduling overhead increases when the windows become smaller, since each subsystem must be processed at each iteration in every window. If the windows are made very small, timesteps chosen to calculate the waveforms are limited by the window size rather than by the local truncation error, and the advantages of the multirate nature of WR will be lost.

The lower bound for the region over which WR contracts uniformly given in Theorem 4.1 is too conservative in most cases to be of direct practical use. As mentioned above, in order for the WR algorithm to be efficient it is important to pick the largest windows over which the iterations actually contract uniformly, but the theorem only provides a worst-case estimate. Since it is difficult to determine *a priori* a reasonable window size to use for a given nonlinear problem, window sizes are usually determined dynamically, by monitoring the computed iterations(See Chapter 7)[18]. Since Theorem 5.2 guarantees the convergence of WR over *any* finite interval, a dynamic scheme does not have to pick the window sizes very accurately. The only cost of a bad choice of window is loss of efficiency; the relaxation *still* converges.

SECTION 5.2 - PARTITIONING LARGE SYSTEMS

In Algorithm 4.1, the system equations are solved as single differential equations in one unknown, and these solutions are iterated until convergence. If this kind of node-by-node decomposition strategy is used for systems with even just a few tightly coupled nodes, the WR algorithm will converge very slowly. As an example, consider the three-node circuit in Fig. 5.3a, a two-inverter chain separated by a resistor-capacitor network. In this case, the resistor-capacitor network is intended to model wiring delays, so the resistor has a large conductance compared to the other conductances in the circuit. The current equations for the system can be written down by inspection and are

$$C\dot{x}_1 + i_{m1}(x_1, vdd) + i_{m2}(x_1, u) + g(x_1 - x_2) = 0 \qquad [5.20]$$

$$C\dot{x}_2 \, g \, (x_2 - x_1) = 0$$

$$C\dot{x}_3 \, i_{m3}(x_3, x_2) + i_{m4}(x_3, vdd) = 0.$$

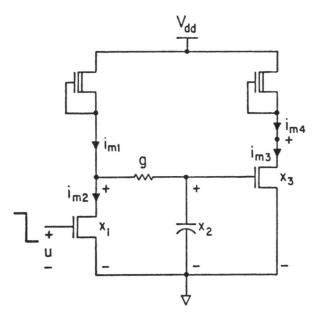

Figure 5.3a - Inverter with Delay

Linearizing and normalizing time (so that the simulation interval $[0,T]$ is converted to $[0,1]$) yields a 3x3 linear equation:

$$\begin{bmatrix} \dot{x}_1 \\ \dot{x}_2 \\ \dot{x}_3 \end{bmatrix} = \begin{bmatrix} -10 & 9.5 & 0 \\ 9.5 & -9.5 & 0 \\ 0 & -1 & -1 \end{bmatrix} \begin{bmatrix} x_1 \\ x_2 \\ x_3 \end{bmatrix} + \begin{bmatrix} 5 \\ 0 \\ 0 \end{bmatrix} \qquad [5.21]$$

$$x_1(0) = x_2(0) = 0 \quad x_3(0) = 5.$$

Algorithm 4.1 was used to solve the original nonlinear system. The input $u(t)$, the exact solution for x_2, and the first, fifth and tenth iteration waveforms generated by the WR algorithm for x_2 are plotted in Fig. 5.3b. As the plot indicates, the iteration waveforms for this example are converging very slowly. The reason for this slow convergence can be seen by examining the linearized system. It is clear x_1 and x_2 are tightly coupled by the small resistor modeling the wiring delay.

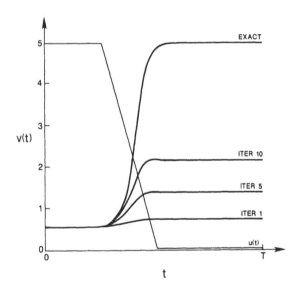

Figure 5.3b – WR Iterations from Inverter with Delay

If Algorithm 4.1 is modified so that x_1 and x_2 are lumped together and solved directly, we get the following iteration equations:

$$\begin{bmatrix} \dot{x}_1^{k+1} \\ \dot{x}_2^{k+1} \end{bmatrix} = \begin{bmatrix} -10 & 9.5 & 0 \\ 9.5 & -9.5 & 0 \end{bmatrix} \begin{bmatrix} x_1^{k+1} \\ x_2^{k+1} \\ x_3^k \end{bmatrix} + \begin{bmatrix} 5 \\ 0 \end{bmatrix} \qquad [5.22]$$

$$\dot{x}_3^{k+1} = -x_2^{k+1} - x_3^{k+1}.$$

The modified WR algorithm now converges in one iteration, because x_3 depends only on the "block" of x_1 and x_2, and that block is independent of x_3 .

As the example above shows, lumping together tightly coupled nodes and solving them directly can greatly improve the efficiency of the WR algorithm. For this reason, the first step in almost every WR-based program is to *partition* the system: to scan all the nodes in the system and determine which should be lumped together and solved directly. Partitioning "well" is difficult for several reasons. If too many nodes are lumped together, the advantages of using relaxation will be lost, but if any tightly coupled nodes are not lumped together then the WR algorithm will converge very slowly. And since the aim of WR is to perform the simulation rapidly, it is important that the partitioning step not be computationally expensive.

Several approaches have been applied to solve this partitioning problem. The first approach is to require the user to partition the system[15]. This technique is reasonable for the simulation of large digital integrated circuits because usually the large circuit has already been broken up into small, fairly independent pieces to make the design easier to understand and manage. However, what is a sensible partitioning from a design point of view may not be a good partitioning for the WR algorithm. For this reason programs that require the user to partition the system sometimes perform a "sanity check" on the partitioning. A warning is issued if there are tightly coupled nodes that have not been lumped together.

A second approach to partitioning, also tailored to digital integrated circuits, is the functional-extraction method[16]. In this method the equations that describe the system are carefully examined to try to find functional blocks (i.e. a *nand* gate or a *flip-flop*). It is then assumed that nodes of the system that are members of the same functional block are tightly coupled, and are therefore grouped together. This type of partitioning is difficult to perform, since the algorithm must recognize broad classes of functional blocks, or nonstandard blocks may not be treated properly. However, the functional-extraction method can produce very good partitions because the relative importance of the coupling of the nodes can be accurately estimated.

Since it is the intent of the partitioning to improve the speed of convergence of the relaxation, it is sensible to partition a large circuit with this speed, rather than topology or functionality, in mind. In this section we will develop an algorithm based on this idea. As it is difficult to get estimates of the speed of WR convergence directly, we will start with an exact analysis of a relaxation algorithm applied to a simple 2x2 linear algebraic example, and then lift the result to a heuristic for partitioning large linear algebraic problems. Then a relationship will be established between the convergence speed of the linear WR algorithm, and that of two linear algebraic problems.

The following definition will be useful for describing the rate of convergence of relaxation algorithms.

<u>Definition 5.3:</u> Let $x^k \in \mathbb{R}^n$ be generated by the kth iteration of an algebraic relaxation algorithm applied to a system of the form $f(x) = 0$, where $x \in \mathbb{R}^n$ and $f:\mathbb{R}^n \to \mathbb{R}^n$. Then the iteration factor γ is defined as the smallest positive number such that

$$\| x^{k+1} - x^k \| \leq \gamma \| x^k - x^{k-1} \|$$

for any $k > 0$, and any bounded initial guess x^0 ∎.

Since the difference between the exact solution, x, and the result of the kth step of a relaxation, x^k, is less than $(\gamma)^k \|x - x^0\|$, the size of γ is an indication of how fast the relaxation converges. If γ is much less than 1 then the relaxation is certain to converge rapidly, but if $\gamma \geq 1$ the relaxation may not converge, and if γ is close to 1 the convergence may be very slow.

Consider the simple 2x2 matrix problem,

$$\begin{bmatrix} a_{11} & a_{12} \\ a_{21} & a_{22} \end{bmatrix} \begin{bmatrix} x_1 \\ x_2 \end{bmatrix} = \begin{bmatrix} b_1 \\ b_2 \end{bmatrix} \qquad [5.23]$$

If the Gauss-Jacobi relaxation algorithm is used to solve Eqn. (5.23) (See Section 3.2) then the iteration factor is bounded below by the spectral radius of

$$\begin{bmatrix} 0 & \dfrac{a_{12}}{a_{11}} \\ \dfrac{a_{21}}{a_{22}} & 0 \end{bmatrix} \qquad [5.24a]$$

which is

$$\gamma^{gj} = \left| \sqrt{\dfrac{a_{21}a_{12}}{a_{11}a_{22}}} \right| \qquad [5.24b]$$

and if the Gauss-Seidel relaxation algorithm is used, then the iteration factor is bounded below by the spectral radius of

$$\begin{bmatrix} 0 & \dfrac{a_{12}}{a_{11}} \\ 0 & \dfrac{a_{21}a_{12}}{a_{11}a_{22}} \end{bmatrix}, \qquad [5.25a]$$

which is

$$\gamma^{gs} = \left| \dfrac{a_{21}a_{12}}{a_{11}a_{22}} \right|. \qquad [5.25b]$$

For the 2x2 linear system of Eqn. (5.23), Eqn. (5.24b) and Eqn. (5.25b) can be used to decide whether to use relaxation or to lump the two nodes together and use direct methods. The criterion that γ^{g} be small (much less than one), which we will refer to as the *diagonally dominant loop* criterion, has proved to be a useful heuristic for partitioning the large systems generated by circuit problems. For the linear algebraic problem

$$Ax = b \qquad [5.26]$$

where $x, b \in \mathbb{R}^n$; $A \in \mathbb{R}^{n \times n}$ and is invertible; $A = (a_{ij})$, we have the following partitioning algorithm.

Algorithm 5.1 Diagonally Dominant Loop Partitioning for $Ax = b$

    for all (i,j in N) {
        if ($\dfrac{a_i \rho_{ji}}{a_j \rho_{ii}} > \alpha$) { x_i is lumped with x_j }
    }

■

The constant α depends on the problem, and is roughly related to the desired iteration factor, so the smaller α is made, the more likely that nodes will be lumped together.

Although Algorithm 5.1 works well for the matrices generated by a wide variety of circuit problems, it is only a heuristic. There are circuit examples for which the diagonally dominant loop criterion does not indicate tightly coupled nodes that should be placed in the same partition. A particularly common circuit example for which Algorithm 5.1 does not lump tightly coupled nodes together is given in Fig. 5.4, an inverter driving a series of resistors. This is just a more complex version of the example given at the beginning of this section. The KCL equations for the circuit, approximating the inverter's output as a one-volt voltage source, are

$$0.01x_1 + 10.0(x_1 - x_2) = 0.01$$

$$10.0(x_2 - x_1) + 1.0(x_2 - x_3) = 0$$

$$1.0(x_3 - x_2) + 10.0(x_3 - x_4) = 0$$

$$0.01x_4 + 10.0(x_4 - x_3) = 0$$

or in matrix form,

$$
\begin{bmatrix}
10.01 & -10.0 & 0.0 & 0.0 \\
-10.0 & 11.0 & -1.0 & 0.0 \\
0.0 & -1.0 & 11.0 & -10.0 \\
0.0 & 0.0 & -10.0 & 10.01
\end{bmatrix}
\begin{bmatrix}
x_1 \\
x_2 \\
x_3 \\
x_4
\end{bmatrix}
=
\begin{bmatrix}
1.0 \\
0 \\
0 \\
0
\end{bmatrix} .
\qquad [5.27]
$$

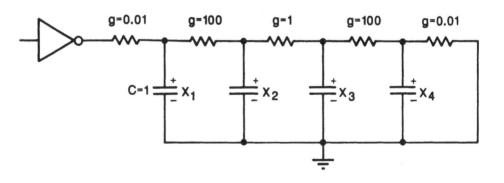

Figure 5.4 - Inverter Driving a Series of Resistors

If Algorithm 5.1 is used to partition the matrix in Eqn. (5.27) and $\alpha = 0.1$, then x_1 will be lumped with x_2, and x_3 will be lumped with x_4. The spectral radius for the iteration matrix generated by applying block Gauss-Seidel relaxation to the partitioned subsystems is $\simeq 0.98$. Since the spectral radius is very close to one, γ^{gi} will be close to one or larger, and the relaxation will converge slowly.

The reason the diagonally dominant loop criterion sometimes produces misleading results is that it is too local a criterion; it only indicates how mutually coupled two nodes are, compared to how coupled they are to other nodes in the problem. If two nodes are extremely tightly coupled as are the pairs x_1, x_2 and x_3, x_4 in the example of Eqn. (5.27), then each of the nodes in the pair will appear relatively loosely coupled to other nodes in the problem, even if they are tightly enough coupled to other nodes to slow the relaxation.

It is possible to modify the diagonally dominant loop partitioning algorithm so that it will produce good partitions for problems which contain subsystems like the example of Eqn. (5.27). To demonstrate the algorithm, we consider a different approach to partitioning. Consider a problem of the form of Eqn. (5.26), $Ax - b = 0$, and define $\lambda_i = \dfrac{\partial x_i}{\partial b_i}$, which is just the ith diagonal term of of A^{-1}. Then a new algorithm is generated by replacing $\dfrac{1}{a_{ii}}$ with λ_{ii} in Algorithm 5.1.

<u>Algorithm 5.2 - Reduced System Partitioning for $Ax = b$</u>

     for all (i in n) { compute λ_{ii} }

     for all (i,j in N) {

         if ($a_{ij}a_{ji}\lambda_i\lambda_j > \alpha$) { x_i is lumped with x_j }

     }

■

A simple circuit interpretation can be given for the two partitioning algorithms based on Norton equivalents[36]. Using the diagonally dominant loop criterion directly to decide whether or not to lump node x_2 with x_3 amounts to examining a circuit for which the elements to the right of x_2 and to the left of x_3 have been replaced with a current source in parallel with a 0.1 ohm resistor to ground. Using the reduced system partitioning algorithm amounts to using the exact equivalent for the circuit in Fig. 5.4; that is, to replace the elements to the right of x_2 and to the left of x_3 with their Norton equivalent, a current source in parallel with a 100.1 ohm resistor to ground. The diagonally dominant loop test applied to this reduced system indicates correctly that $\gamma \simeq 0.98$.

Of course, computing the inverse of A is a foolish approach to partitioning if the problem is to compute a matrix solution by relaxation. It is a useful notion, though, because there are many cases where reasonable approximations to λ_i can be computed easily, as we will demonstrate in Chapter 7.

Both the diagonally dominant loop and the reduced system criterion are heuristic techniques for partitioning linear algebraic systems. The next step is to lift the techique to an approach for partitioning linear differential systems of the form of Eqn. (2.10),

$$C\dot{x}(t) \; = \; Ax(t) \; + \; u(t) \quad x(0) \; = \; x_0 \qquad\qquad [5.28]$$

where $C, A \in \mathbb{R}^{n \times n}$; C is nonsingular; and $x(t) \in \mathbb{R}^n$. We will start by presenting the waveform equivalent of the iteration factor.

<u>Definition 5.4:</u> Let $x^k:[0,T] \to \mathbb{R}^n$ be the function generated by the kth iteration of the WR algorithm applied to a system of the form of Eqn. (5.28). Then the WR uniform iteration factor, γ_∞^{WR} , for the system is defined as the smallest positive number such that

$$\max_{[0,T]} \| x^{k+1}(t) - x^k(t) \| \; \leq \; \gamma_\infty^{WR} \; \max_{[0,T]} \| x^k(t) - x^{k-1}(t) \|$$

for any $k > 0$, any continously differentiable initial guess x^0, and any piecewise continuous input u.∎

There are two ways to reduce γ_∞^{WR}. The first, discussed in Section 5.1, is to reduce the simulation interval $[0,T]$ until γ_∞^{WR} is less than one. The second approach is to partition the circuit into loosely coupled subsystems. A combination of the two techniques is needed to allow for reasonably large windows and reasonably small partitions.

As mentioned above, it is difficult to estimate γ_{∞}^{WR} directly for a given problem of the form of Eqn. (5.28). There are the following theorems which relate γ_{∞}^{WR} to iteration factors resulting from a simplified system of equations.

Theorem 5.3: Let γ_{∞}^{WR} be the WR uniform iteration factor for a given system of equations of the form of Eqn. (5.28) solved on [0,T]. Then in the limit as $T \to \infty$, γ_{∞}^{WR} is bounded below by the spectral radius of $(L_a + D_a)^{-1}U_a$ where L_a, D_a, and U_a are the strictly lower-triangular, diagonal, and strictly upper-triangular portions of A given in Eqn. (5.28). ∎

The theorem is simple to prove given the following lemma, the proof of which is given in [32].

Lemma 5.2: Let F be any linear map such that $y = Fx, y, x:[0, \infty) \to \mathbb{R}^n$. Define $y(s), x(s), F(s)$ as the Laplace tranforms of y, x, and F respectively. Then the spectral radius of the map F, $\rho(F)$ is equal to the max, $\rho(F(s))$. ∎

Proof of Theorem 5.3

Let $L_c, D_c, \wedge U_c$ be the strictly lower-triangular, diagonal, and upper-triangular portions of C respectively. Similarly, let $L_a, D_a, \wedge U_a$ be the strictly lower-triangular, diagonal, and upper-triangular portions of A. Using this notation, the Gauss-Seidel WR iteration equation applied to solving Eqn. (5.28) is

$$(L_c + D_c)\dot{x}^{k+1}(t) + U_c\dot{x}^k(t) = (L_a + D_a)x^{k+1}(t) + U_a x^k(t). \qquad [5.29]$$

Define $\varepsilon^k(t) = x^k(t) - x^{k-1}(t)$. Taking the difference between the $(k + 1)^{st}$ and k^{th} iterations of Eqn. (5.29) yields

$$(L_c + D_c)\dot{\varepsilon}^{k+1}(t) + U_c\dot{\varepsilon}^k(t) = (L_a + D_a)\varepsilon^{k+1}(t) + U_a\varepsilon^k(t). \qquad [5.30]$$

Taking the Laplace transform of Eqn. (5.30) yields

$$s(L_c + D_c)\varepsilon^{k+1}(s) + sU_c\varepsilon^k(s) = (L_a + D_a)\varepsilon^{k+1}(s) + U_a\varepsilon^k(s). \quad [5.31]$$

Reorganizing this equation, and assuming the diagonal elements of C are nonzero, gives

$$\varepsilon^{k+1}(s) = [s(L_c + D_c) + (L_a + D_a)]^{-1}(sU_c + U_a)\varepsilon^k(s). \quad [5.32]$$

Eqn. (5.32) is in the from for which Lemma 5.2 can be applied. In particular, the linear map $F(s)$ in that lemma for this case is

$$F(s) = [s(L_c + D_c) + (L_a + D_a)]^{-1}(sU_c + U_a). \quad [5.33]$$

The theorem then follows from the fact that

$$F(0) = [L_a + D_a]^{-1}U_a \quad [5.34]$$

and from Lemma 5.2. ∎

Theorem 5.4: Let γ_{∞}^{WR} be the WR uniform iteration factor for a given system of equations of the form of Eqn. (5.28). Then γ_{∞}^{WR} is is bounded below by the spectral radius of $(L_c + D_c)^{-1}U_c$ where $L_c, D_c, \wedge U_c$ are the strictly lower-triangular, diagonal, and strictly upper-triangular portions of C given in Eqn. (5.28). ∎

Proof of Theorem 5.4

    The proof of Theorem 5.4 follows trivially from Eqn. (5.33), again applying Lemma 5.2 and evaluating $F(s)$ in the limit as $s \to \infty$. In this case there is also a time-domain proof, and we present it for completeness.

    Algebraically reorganizing Eqn. (5.30),

$$\dot{\varepsilon}^{k+1}(t) = -(L_c + D_c)^{-1}U_c\dot{\varepsilon}^k(t) - \quad [5.35]$$

$$(L_c + D_c)^{-1}(L_a + D_a)\varepsilon^{k+1}(t) \ + \ (L_c + D_c)^{-1}U_a\varepsilon^k(t).$$

Integrating Eqn. (5.35) and using the fact that $\varepsilon(0) = 0$,

$$\varepsilon^{k+1}(t) \ = \ -(L_c + D_c)^{-1}U_c\varepsilon^k(t) \ -$$ [5.36]

$$\int_0^t (L_c + D_c)^{-1}(L_a + D_a)\varepsilon^{k+1}(\tau)d\tau \ + \ \int_0^t (L_c + D_c)^{-1}U_a\varepsilon^k(\tau)d\tau.$$

Since Eqn. (5.36) holds for all t, it holds as $t \to 0$, which proves the theorem. ∎

In Eqn. (5.28), A and C represent the matrix of linear conductors and linear capacitors, respectively. The two theorems above indicate that it is possible to get lower-bound estimates of γ_∞^{WR} by examining circuits where only the capacitances and conductances are independently present. These estimates are lower bounds; hence, to decrease γ_∞^{WR} below a desired α, it is *necessary* to partition in such a way that the iteration factors for the Gauss-Seidel iteration applied to the algebraic systems are decreased below α.

SECTION 5.3 - ORDERING THE EQUATIONS

When applying the Gauss-Seidel WR algorithm to a decomposed system of differential equations, the order in which the equations are solved can strongly affect the number of WR iterations required to achieve satisfactory convergence. In order to explain this effect, consider using the Gauss-Seidel relaxation algorithm to solve a large system of linear algebraic equations of the form

$$Ax = b$$

where $x,b \in \mathbb{R}^n$ and $A \in \mathbb{R}^{n \times n}$. As was shown in Chapter 3, the Gauss-Seidel algorithm can be written in matrix form as

$$(L + D)x^k \ + \ Ux^{k-1} = b$$ [5.37]

where $L, D, U \in \mathbb{R}^{n \times n}$ are strictly lower triangular, diagonal, and strictly upper triangular respectively, and are such that $A = L + D + U$. Taking the difference between the kth and $(k - 1)^{st}$ iterations we get

$$x^k - x^{k-1} = (L + D)^{-1} U(x^{k-1} - x^{k-2})$$ [5.38]

assuming $L + D$ is nonsingular (i.e., the entries of D are nonzero). Taking the norm of both sides leads to

$$\| x^k - x^{k-1} \| \leq \| (L + D)^{-1} U \| \, \| (x^{k-1} - x^{k-2}) \|.$$ [5.39]

Clearly, the closer $\| (L + D)^{-1} U \|$ is to zero, the faster the relaxation will converge.

For example, suppose that A is lower triangular. Then $U = 0$ and $\| (L + D)^{-1} U \| = 0$ and therefore the relaxation converges in one iteration. However, the order of the single equations represented by the rows of A is not unique. The equations could be reverse ordered, so that b_i becomes b_{n+1-i} and $Ax = b$ becomes $\tilde{A} \tilde{x} = \tilde{b}$ where $\tilde{A}$ is a row permutation of A. In this case, $\tilde{U}$ will not be zero, $\| (\tilde{L} + \tilde{D})^{-1} \tilde{U} \| \neq 0$ and therefore the relaxation is not likely to converge in one iteration.

This is, of course, an extreme example, but it does indicate that if the Gauss-Seidel relaxation algorithm is used, it is possible to reorder the system so that the number of iterations required to achieve convergence can be significantly reduced. In particular, a reordering should attempt to move as many of the large off-diagonal elements of the matrix as possible into the lower-triangular portion.

As discussed in Section 5.2, subsets of nodes in a large circuit may be mutually tightly coupled, and in order to insure that the relaxation algorithm converges rapidly when applied to such a circuit, these subsets are grouped together into subcircuits and solved with direct methods. This corresponds to a *block* relaxation method, and an ordering algorithm applied to a system being solved with block relaxation should attempt to make the problem as *block lower triangular* as possible.

In some sense, partitioning and ordering the subsystem of equations are performing similar functions. They are both attempting to eliminate slow relaxation convergence due to two nodes in a large circuit being tightly coupled. There is, however, a *key* difference. If, for example, x_i is strongly dependent on x_j and x_j is strongly dependent on x_i , then a partitioning algorithm should lump the two nodes together into one subsystem. However, if x_i is strongly dependent on x_j, but x_j is *weakly* dependent on x_i then node i and node j should not be lumped together, but the ordering algorithm should insure that the system is block lower triangular by ordering the equations so that x_j is computed before computing x_i.

Resistors and capacitors do not exhibit the kind of unidirectional coupling that is of concern to the ordering algorithm. In fact, the only element type of concern to the ordering algorithm the transistor, because it exhibits unidirectional coupling. That is, the drain and source terminals of an MOS transistor are strongly dependent on the gate terminal of the transistor, but the gate is almost independent of the drain and source. Clearly, this implies that the subsystems containing the given transistor's drain or source should be analyzed after the subsystem containing the given transistor's gate. We will return to this point in Chapter 7, where we will describe an ordering algorithm for MOS circuits in detail.

CHAPTER 6 - DISCRETIZED WR ALGORITHMS

To compute the iteration waveforms for the WR algorithm it is usually necessary to solve systems of nonlinear ordinary differential equations. If multistep integration formulas are used to solve for the iteration waveforms, the differential equations that describe the decomposed systems will not be solved exactly. Therefore, the convergence theorem presented in Section 4.2 does not guarantee the convergence of this *discretized* WR algorithm. However, the discretized WR algorithm is a nonstationary method and the therefore the theorem presented in Section 4.4 applies, and guarantees WR convergence to the solution of the given system of ODE's when the *global* discretization error is driven to zero with the WR iterations. Reducing the error with the iteration is also a reasonable practical approach to insuring the convergence of the WR algorithm under discretizations. Timesteps for numerical integration methods are usually chosen to insure that estimates of the local truncation error are kept below some supplied criteria. Reducing these criteria as relaxation iterations progress insures that the WR algorithm will converge.

Viewing the discretized WR algorithm as a nonstationary method, although simple and practical, lends no insight into why the discretized WR algorithm may not converge in some cases, and therefore provides no guidance for selecting a numerical integration method. It also does not allow for comparison to more classical integration methods. For this reason the interaction between WR algorithms and multistep integration methods will be considered in detail in this chapter. In the first section, the discretized WR algorithm will be analyzed assuming that every differential equation in the system is discretized identically (hereafter referred to as the *global-timestep* case). A simple example will be presented that demonstrates a possible breakdown of the WR method under discretizations. The nonconvergence will be investigated by comparing the global-timestep discretized WR algorithm to the relaxation-Newton methods of Section 3.3. The major result is the proof of a strong comparison theorem for linear systems: the global timesteps required to insure WR convergence are identical to the timesteps required to insure convergence of the relaxation methods presented in Section 3.3.

In the second section of this chapter we will investigate the fixed-global-timestep discretized WR algorithm. We will demonstrate by example the difficulty in extending the convergence theorems for the algebraic relaxtion algorithms to the discretized WR algorithm, and present a convergence proof for the discretized WR algorithm for the fixed-global-timestep case. In the third section, the global-timestep restriction will be lifted, and a theorem demonstrating the convergence of the multirate timestep case for systems in normal form will be presented.

SECTION 6.1 - THE GLOBAL-TIMESTEP CASE

Consider the two-node inverter circuit in Fig. 6.1. The current equations at each node can be written by inspection, and are:

$$C\dot{x}_1 + g_1 x_1 + g_2(x_1 - x_2) = 0 \qquad [6.1]$$

$$C\dot{x}_2 + g_2(x_2 - x_1) + i_{m1}(x_1, x_2) + i_{m2}(x_1) = 0$$

$$x_1(0) = x_2(0) = 0.$$

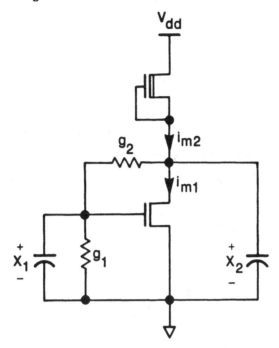

Figure 6.1 - Two Node Inverter Circuit

In order to generate a simple linear example, i_{m1}, i_{m2} were linearized about the point where the input and output voltages were equal to half of the supply voltage. Time is normalized to seconds to obtain the following 2x2 example:

$$\dot{x}_1 = -x_1 + 0.1x_2 \qquad [6.2]$$

$$\dot{x}_2 = -\lambda x_1 + -x_2$$

$$x_1(0) = x_2(0) = 0.$$

Note that the initial conditions given for the above example identify a stable equilibrium point.

The Gauss-Seidel WR iteration equations for the linear system example are

$$\dot{x}_1^{k+1} = -x_1^{k+1} + 0.1x_2^k \qquad [6.3]$$

$$\dot{x}_2^{k+1} = -\lambda x_1^{k+1} - x_2^{k+1}$$

$$x_1^k(0) \ = \ x_2^k(0) \ = \ 0 \textit{for } \forall k.$$

Applying the implicit-Euler numerical integration method with a fixed timestep h, (that is, approximating $\dot{x}(nh)$ by $\frac{1}{h}[x(nh) - x((n-1)h)]$) to solve the decomposed equations yields the following recursion equation for $x_2^k(n)$:

$$x_2^{k+1}(n) \ = \ \frac{1}{1+h} x_2^{k+1}(n-1) \ - \qquad\qquad\qquad [6.4]$$

$$\frac{\lambda h}{(1+h)^2} \ [\frac{x_1(0)}{(1+h)^n} \ + \ 0.1h \sum_{j=1}^{n}(1+h)^{j-n}x_2^k(j)].$$

For example, let $\lambda = 200$, $h = 0.5$, and as an initial guess use $x_2^0(nh) = nh$, which is far from the exact solution $x_2^0(nh) = 0$. The computed sequences for the initial guess and the first, second and third iterations of Eqn. (6.4) are presented in Table 6.1.

TABLE 6.1 - IMPLICIT-EULER COMPUTED WR ITERATIONS					
STEP	TIME	INITIAL	ITER #1	ITER #2	ITER #3
0	0	0	0	0	0
1	0.5	0.5	-1.111	2.469	-6.487
2	1.0	1.0	-3.704	152	-32.92
3	1.5	1.5	-7.778	355	-111.6
4	2.0	2.0	-13.17	66.21	-281.3
5	2.5	2.5	-19.66	117.9	-587.5
6	3.0	3.0	-27.02	187.9	-1075
7	3.5	3.5	-36.07	276.0	-1786
8	4.0	4.0	-43.64	385	-2751
9	4.5	4.5	-52.60	502.9	-3992
10	5.0	5.0	-61.85	638.4	-5519

As Table 6.1 indicates, the WR algorithm diverges for this example. In fact, Eqn. (6.4) indicates that the WR algorithm will converge only if

$$\frac{h}{(1 + h)} < \frac{1}{\sqrt{0.1\lambda}}.$$ [6.5]

The constraints on the timesteps for which the global-timestep discretized WR algorithm will converge are very similar to the constraints on the timesteps for which the relaxation-Newton algorithm applied to Eqn. (3.5) will converge(see Section 3.3). In fact, for linear problems there is the following comparison theorem.

Theorem 6.1: Let a consistent and stable multistep integration algorithm be applied to a linear system of the form

$$C\dot{x}(t) = Ax(t) \qquad x(0) = x_0$$ [6.6]

where $C, A \in \mathbb{R}^{n \times n}$; C is nonsingular; and $x(t) \in \mathbb{R}^n$. Assume further that the Gauss-Seidel(Jacobi) algebraic relaxation algorithm is used to solve the linear algebraic equations generated by the integration algorithm (as described in Section 3.3). Given a sequence of timesteps, $\{h_m\}$, the Gauss-Seidel(Jacobi) algebraic relaxation algorithm will converge at every step, for any initial guess, if and only if the global-timestep discretized Gauss-Seidel(Jacobi) WR algorithm, generated by solving the iteration equations with the same multistep integration algorithm and same timestep sequence, converges for any initial guess. ∎

Proof of Theorem 6.1

The algebraic equation generated by applying a multistep integration algorithm to Eqn. (6.6) is

$$\sum_{i=0}^{k} \alpha_i C\hat{x}(\tau_{m-i}) = h_m \sum_{i=0}^{l} \beta_i A\hat{x}(\tau_{m-i}),$$ [6.7]

or reorganizing,

$$[C - h_m \beta_0 A] \hat{x}(\tau_m) + \sum_{i=1}^{k} \alpha_i C \hat{x}(\tau_{m-i}) - h_m \sum_{i=1}^{l} \beta_i A \hat{x}(\tau_{m-i}) = 0. \quad [6.8]$$

Let L_c, D_c, U_c be the strictly lower-triangular, diagonal, and upper-triangular portions of C. Similarly, let L_a, D_a, U_a be the strictly lower-triangular, diagonal, and upper-triangular portions of A. Using this notation, the Gauss-Seidel relaxation iteration equation applied to solving Eqn. (6.8) for $x(\tau_m)$ is

$$[(L_c + D_c) - h_m \beta_0 (L_a + D_a)] \hat{x}^k(\tau_m) + [U_c - h_m \beta_0 U_a] \hat{x}^{k-1}(\tau_m) +$$

$$\sum_{i=1}^{k} \alpha_i C \hat{x}(\tau_{m-i}) - h_m \sum_{i=1}^{l} \beta_i A \hat{x}(\tau_{m-i}) = 0.$$

Taking the difference between the kth and $(k-1)^{st}$ iterations and substituting $\delta^k(\tau_m)$ for $x^k(\tau_m) - x^{k-1}(\tau_m)$ leads to

$$[(L_c + D_c) - h_m \beta_0 (L_a + D_a)] \delta^k(\tau_m) = - [U_c - h_m \beta_0 U_a] \delta^{k-1}(\tau_m), \quad [6.9]$$

from which it follows that the relaxation will converge at the mth timepoint for any initial guess if and only if the spectral radius of

$$[(L_c + D_c) - h_m \beta_0 (L_a + D_a)]^{-1} [U_c - h_m \beta_0 U_a] \quad [6.10]$$

is less than one.

If the Gauss-Seidel WR algorithm is used to solve Eqn. (6.6), the iteration equation for $x(t)$ is (using the above notation)

$$(L_c + D_c) \dot{x}^k(t) + U_c \dot{x}^{k-1}(t) = (L_a + D_a) x^k(t) + U_a x^{k-1}(t). \quad [6.11]$$

Applying the multistep integration algorithm to solve Eqn. (6.11) for x^{k+1} yields

$$[(L_c + D_c) - h_m\beta_0(L_a + D_a)]\hat{x}^k(\tau_m) + [U_c - h_m\beta_0 U_a]\hat{x}^{k-1}(\tau_m) + \qquad [6.12]$$

$$\sum_{i=1}^{k} \alpha_i[(L_c + D_c)\hat{x}^k(\tau_{m-i}) + U_c\hat{x}^{k-1}(\tau_{m-i})] -$$

$$h_m\sum_{i=1}^{l} \beta_i[(L_a + D_a)\hat{x}^k(\tau_{m-i}) + U_a\hat{x}^{k-1}(\tau_{m-i})] = 0.$$

Taking the difference between the kth and $(k-1)^{st}$ iterations and substituting $\delta^k(\tau_m)$ for $x^k(\tau_m) - x^{k-1}(\tau_m)$ leads to

$$[(L_c + D_c) - h_m\beta_0(L_a + D_a)]\delta^k(\tau_m) + [U_c - h_m\beta_0 U_a]\delta^{k-1}(\tau_m) + \qquad [6.13]$$

$$\sum_{i=1}^{k} \alpha_i[(L_c + D_c)\delta^k(\tau_{m-i}) + U_c\delta^{k-1}(\tau_{m-i})] -$$

$$h_m\sum_{i=1}^{l} \beta_i[(L_a + D_a)\delta^k(\tau_{m-i}) + U_a\delta^{k-1}(\tau_{m-i}))] = 0.$$

To show that the discretized WR algorithm will only converge if the algebraic re-laxation converges, let l be a timestep for which the spectral radius of the matrix in Eqn. (6.10) is not less than one. Use as an initial guess any sequence for which the first $l-1$ points are the exact solution to the discretized problem. Then $\delta^k(\tau_m) = 0$ for $m < l$, and Eqn. (6.13) is again identical to Eqn. (6.9), and is not convergent.

An inductive argument is used to prove that if the algebraic relaxation is convergent then the discretized WR algorithm is convergent. Assume that the theorem holds for $m < l$; then, $\delta^k(\tau_{l-1})$ will go to zero as $k \to \infty$. As this occurs, Eqn. (6.13) for the lth step converges to Eqn. (6.9). The algebraic relaxation converges and therefore the spectral radius of the matrix in Eqn. (6.10) for the lth step is less than one. This implies that Eqn. (6.9) represents a contraction mapping in some norm at the lth step, and the contraction mapping theorem (Section 4.2) can be applied to guarantee that Eqn. (6.13) converges at the lth step. Note that $\delta^k(\tau_m) = 0$ for all $m \leq 0$, and therefore Eqn. (6.13) is identical to Eqn. (6.9) for $m = 1$ which completes the induction.∎

The above theorem holds nonlinear systems of the form of Eqn. (2.10) if it is assumed that an arbitrarily close initial guess for each of the relaxation schemes is available. Although this is not a realistic assumption, it does indicate that even for nonlinear systems the two algorithms present very similar timestep constraints for a numerical integration method.

SECTION 6.2 - FIXED GLOBAL-TIMESTEP WR CONVERGENCE THEOREM

It is possible to generalize the proof of Theorem 6.1 to a proof for the global-timestep discretized WR algorithm for nonlinear problems (but, as mentioned above, the comparison to the relaxation-Newton methods would no longer hold). A different approach will be taken, because the approach followed in Theorem 6.1 does not prove that the discretized WR algorithm converges on a fixed time interval as the timesteps become small.

To illustrate this point by example, consider solving Eqn. (6.3) using the explicit-Euler method. The recursion equation for the $x_2^k(n)$'s is:

$$x_2^{k+1}(n+1) = (1-h)x_2^{k+1}(n) - 0.1\lambda h^2[(1-h)^n x_1^k(0) + \sum_{j=1}^{n-1}(1-h)^{n-1-j}x_2^k(j)$$

The computed sequences $\{x_2^{k+1}\}$'s for the initial guess and the first, second and third iterations of the above equation are given in Table 6.2, for the case of $\lambda = 200$, $h = 0.5$ and $x_2^0(nh) = nh$.

TABLE 6.2 – EXPLICIT-EULER COMPUTED WR ITERATIONS					
STEP	TIME	INITIAL	ITER #1	ITER #2	ITER #3
0	0	0	0	0	0
1	0.5	0.5	0	0	0
2	1.0	1.0	0	0	0
3	1.5	1.5	-0.625	0	0
4	2.0	2.0	-1.875	0	0
5	2.5	2.5	3.594	0.7813	0
6	3.0	3.0	-6.625	3.125	0
7	3.5	3.5	-7.852	7.422	-0.977
8	4.0	4.0	-10.19	13.67	-4.883
9	4.5	4.5	-12.61	21.63	-13.92
10	5.0	5.0	-16.06	30.96	-29.79

As the table indicates, the explicit-Euler discretized WR algorithm con-verges in just the manner predicted by Theorem 6.1, a timestep (or two) with each iteration. If in the same example, half the timestep is used, similar results are achieved. That is, the relaxation converges two steps with each iteration. If this were the case no matter how small the timesteps became, that is, each relaxation iteration resulted in only two more timesteps converging, then given a fixed in-terval of interest, the WR algorithm would *not* converge in the limit as the timesteps approached zero. This is not the case for this example, or in general for the discretized WR algorithm. If, for example, $h = 0.05$ then the relaxation converges in a more uniform manner, where the value at *each* timestep rapidly approaches its limit point.

In Section 4.2, the WR algorithm was shown to be a contraction mapping, specifically

$$\max_{[0,T]}^{-bt} \| \dot{x}^k(t) - \dot{x}^l(t) \| \leq \gamma \max_{[0,T]} e^{-bt} \| \dot{x}^{k-1}(t) - x^{l-1}(t) \|$$

where $\gamma, b \in \mathbb{R}$ are dependent on the problem, and $\gamma < 1$. If T is chosen small enough, then $\gamma e^{bT} = \hat{\gamma} < 1$ and the norm becomes

$$\max_{[0,T]} \| \dot{x}^k(t) - \dot{x}^l(t) \| \leq \hat{\gamma} \max_{[0,T]} \| \dot{x}^{k-1}(t) - \dot{x}^{l-1}(t) \|.$$

That is, the WR algorithm converges uniformly over small time intervals (as was discussed in Section 5.2). The next theorem will be an analogous proof for the discretized case. It will be shown that the fixed global-timestep discretized WR algorithm is a contraction in a b norm[29].

Formally, demonstrating that the discretized WR is a contraction in a b norm implies that the discretized WR algorithm converges because of the contraction mapping theorem. Intuitively, the convergence of the discretized WR algorithm in a b norm implies that the discretized WR algorithm has some underlying uniformity in its convergence. This is sufficient to guarantee convergence of the algorithm over a fixed time interval as the timesteps shrink to zero. This is the distinction between Theorem 6.1 and the next theorem.

Theorem 6.2: If, in addition to the assumptions of Theorem 4.1, f in Eqn. (2.10) is differentiable, and the WR iteration equations are solved using a stable, consistent, multistep integration method with a fixed timestep h, then there exists some h_0 such that the sequences $\{x^k(n)\}$ generated by the Gauss-Seidel or Gauss-Jacobi discretized WR algorithm will converge for all $h_0 > h > 0.$ ■

Before proving Theorem 6.2, some standard notation[1,59] will be presented that will also be used in the next section. The fixed-timestep multistep integration algorithms applied to

$$\dot{x}(t) = f(x(t)) \qquad x(0) = x_0, \qquad [6.14]$$

where $x:[0,T] \to \mathbb{R}^n$ and $f:\mathbb{R}^n \to \mathbb{R}^n$, can be represented by backward shift operators. That is, given

$$\sum_{i=0}^{k} \alpha_i \hat{x}(\tau_{m-i}) = h_m \sum_{i=0}^{l} \beta_i f(\hat{x}(\tau_{m-i})) \qquad [6.15]$$

where $\tau_m - \tau_{m-1} = h_m$, we can define

$$\rho(\hat{x}(\tau_m)) = \sum_{i=0}^{k} \alpha_i \hat{x}(\tau_{m-i}) \qquad [6.16a]$$

and

$$\sigma(f(\hat{x}(\tau_m)) = \sum_{i=0}^{l} \beta_i f(\hat{x}(\tau_{m-i})). \qquad [6.16b]$$

Eqn. (6.15) can then be written compactly as

$$\rho(\hat{x}(\tau_m)) = h_m \sigma(f(\hat{x}(\tau_m)). \qquad [6.17]$$

If it is assumed that the operator ρ can be inverted, i.e. that $\hat{x}(\tau_m)$ can be expressed as a function of the right-hand side, then Eqn. (6.17) can be written in the form

$$\hat{x}(\tau_m) = h_m \rho^{-1} \sigma f(\hat{x}(\tau_m)). \qquad [6.18]$$

When such an inverse of ρ exists, it can be shown that Eqn. (6.18) is equivalent to

$$\hat{x}(\tau_m) = \sum_{j=0}^{m} \gamma_j f(\hat{x}(\tau_{m-j})) + x(0). \qquad [6.19]$$

As an example, consider the implicit-Euler integration method applied to Eqn. (6.14). The usual form for the discrete equations is

$$\hat{x}(\tau_m) - \hat{x}(\tau_{m-1}) = h_m f(\hat{x}(\tau_m)) \qquad [6.20]$$

which is in the form of Eqn. (6.17). The implicit-Euler discrete equations can also
be expressed in the form of Eqn. (6.19),

$$\hat{x}(\tau_m) = \sum_{j=0}^{m} h_{m-j} f(\hat{x}(\tau_{m-j})) + x(0). \qquad [6.21]$$

The solution to Eqn. (6.21) is identical to the solution to Eqn. (6.20). The form
of Eqn. (6.17) for the trapezoidal rule is

$$\hat{x}(\tau_m) - \hat{x}(\tau_{m-1}) = 0.5h_m[f(\hat{x}(\tau_m)) + f(\hat{x}(\tau_{m-1}))], \qquad [6.22]$$

which can also be expressed in the form of Eqn. (6.19) as

$$\hat{x}(\tau_m) = 0.5[\, h_m f(\hat{x}(\tau_m)) + h_1 f(x(0))\,] + \qquad [6.23]$$

$$\sum_{j=1}^{m-1} [\, 0.5(h_{m+1-j} + h_{m-j})\, f(\hat{x}(\tau_{m-j}))\,] + x(0).$$

The following lemma[30], will be the key result used in the course of the
proof of Theorem 6.2.

Lemma 6.1: Let $H(b)$ be the map that represents one iteration of the algebraic
Gauss-Seidel or Gauss-Jacobi relaxation algorithm applied to an equation system
of the form $f(x) - b = 0$, where $x,b \in \mathbb{R}^n$, $f:\mathbb{R}^n \rightarrow \mathbb{R}^n$. If f is such that the
Jacobian of f, $\dfrac{\partial f}{\partial x}$, exists, has a uniformly bounded inverse, and is strictly
diagonally dominant uniformly over all x then $H(b)$ is a contraction mapping in
the l_∞ norm and is a Lipschitz continuous function of b.∎

Proof of Lemma 6.1:

As usual, only the Gauss-Seidel case will be proved. It will be shown that if the Gauss-Seidel relaxation algorithm is used to solve $f(x) - b = 0$, then the map implicitly defined by one iteration of the relaxation, $H(b)$, is such that given two arbitrary points, $x^k, y^l \in \mathbb{R}^n$,

$$\| H(b)x^k - H(b)y^l \|_\infty \leq \gamma \| x^k - y^l \|_\infty \qquad [6.24]$$

where $\gamma < 1$.

Define

$$x^{k,i} = (x_1^k, ..., x_i^k, x_{i+1}^{k-1}, ..., x_n^{k-1})^T. \qquad [6.25]$$

The iteration equation for x_1^{k+1} is implicitly defined by

$$f_1(x_1^{k+1}, x_2^k, ..., x_n^k) - b_1 = 0 \qquad [6.26a]$$

or, using the above notation, $f_1(x^{k+1,1}) - b = 0$. In the same notation, the implicit iteration equation for y_1 is

$$f_1(y^{k+1,1}) - b = 0. \qquad [6.26b]$$

Define the function $\psi(t) = f_1(\lambda x^{k+1,1} + (1 - \lambda)y^{k+1,1}) - b_1$ where $\lambda \in [0,1]$. Clearly, $\psi(0) = \psi(1) = 0$. By Rolle's theorem[35] there exists a $\lambda_0 \in (0,1)$ such that

$$\psi'(\lambda_0) = 0 = \sum_{j=1}^n [\frac{\partial f_1}{\partial x_j} (\lambda_0 x^{k+1,1} + (1 - \lambda_0)y^{k+1,1})(x_j^{k+1,1} - y_j^{k+1,1})]. \qquad [6.27]$$

Reorganizing,

$$\frac{\partial f_1}{\partial x_1}(\lambda x^{k+1,1} + (1 - \lambda_0)y^{k+1,1})(x_1^{k+1} - y_1^{k+1}) = \qquad [6.28]$$

$$-\sum_{j=2}^{n} \frac{\partial f_1}{\partial x_j}(\lambda_0 x^{k+1,1} + (1-\lambda_0)y^{k+1,1})(x_j^k - y_j^k).$$

Dividing Eqn. (6.28) by $\dfrac{\partial f_1}{\partial x_1}$, which is bounded away from zero by the uniform strict diagonal dominance of $\dfrac{\partial f}{\partial x}$, and using the fact that $|x_j^k - y_j^k| \leq \|x^k - y^k\|_\infty$ by definition, we get

$$|x_1^{k+1} - y_1^{k+1}| \leq \qquad\qquad\qquad\qquad\qquad\qquad\qquad [6.29]$$

$$\sum_{j=2}^{n} \left| \frac{\dfrac{\partial f_1}{\partial x_j}(\lambda_0 x^{k+1,1} + (1-\lambda_0)y^{k+1,1})}{\dfrac{\partial f_1}{\partial x_1}(\lambda_0 x^{k+1,i} + (1-\lambda_0)y^{k+1,1})} \right| \ \|x^k - y^k\|_\infty.$$

Using the property of f that the Jacobian is strictly diagonally dominant uniformly in x leads to

$$|x_1^{k+1} - y_1^{k+1}| \leq \gamma_1 \|x^k - y^k\|_\infty$$

where $\gamma_1 < 1$. This proves that

$$|H_1(x^k) - H_1(y^k)| \leq \gamma_1 \|x^k - y^k\|_\infty. \qquad\qquad [6.30]$$

A similar argument can be used to show

$$|H_i(x^k) - H_i(y^k)| \leq \gamma_i \|x^k - y^k\|_\infty$$

where $\gamma_i < 1$. Then, if γ is chosen to be the maximum of the γ_i's, $\gamma < 1$ and

$$\|H(x^k) - H(y^k)\|_\infty \leq \gamma \|x^k - y^k\|_\infty, \qquad\qquad [6.31]$$

which proves the first part of the theorem (for a more detailed proof of the general cases, see [30]).

That H is a Lipschitz continuous function of b can be seen by examining the implicitly defined H_1,

$$f_1(x_1^{k+1}, x_2^k, ..., x_n^k) - b_i = 0,$$

which is solved for x_1^{k+1}. A simple application of the implicit function theorem[35] implies that if $\dfrac{\partial f_1}{\partial x_1}$ is bounded away from zero uniformly in x, then x_1^{k+1} is a Lipschitz continuous function of b_1. That $\dfrac{\partial f_1}{\partial x_1}$ is bounded away from zero follows from the assumption that $\dfrac{\partial f}{\partial x}$ is diagonally dominant and has a uniformly bounded inverse.

The argument can be carried inductively to show that for each i, $H_j, j \leq i$ is a Lipschitz continuous function of b and that therefore $H(b)$ is Lipschitz continuous with respect to b.∎

The formal definition of the β norm for a sequence is given below.

Definition 6.1: Let $\{x(\tau_m)\}$, be the sequence generated by a fixed-timestep numerical integration algorithm. That is, $\tau_m - \tau_{m-1} = h$, the fixed timestep, for all m. The β norm for the sequence is defined as

$$\| \{x(\tau_m)\} \|_\beta = \max_m e^{-\beta h m} \| x(\tau_m) \| \qquad [6.32]$$

where $\beta \in \mathbb{R}$.∎

The following simple lemma will be useful for the proof of Theorem 6.2.

Lemma 6.2: Given an arbitrary sequence, $\{x(\tau_m)\}$, the following inequality holds:

$$\left\| \left\{ \sum_{i=1}^m \gamma_i x(\tau_{m-i}) \right\} \right\|_B \leq M \frac{e^{-Bhm}}{1 - e^{-Bhm}} \| \{x(\tau_m)\} \|_B \qquad [6.33]$$

where $M = \max_i |\gamma_i|.\blacksquare$

Proof of Lemma 6.2:

The proof of Lemma 6.2 follows from a simple algebraic argument. From Definition 6.1,

$$\| \{\sum_{i=1}^{m} \gamma_i x(\tau_{m-i})\} \|_B = \max_m e^{-Bhm} \| \sum_{i=1}^{m} \gamma_i x(\tau_{m-i}) \|. \qquad [6.34]$$

Using the norm properties, the term

$$e^{-Bhm} \| \sum_{i=1}^{m} \gamma_i x(\tau_{m-i}) \| \qquad [6.35]$$

can be bounded by

$$e^{-Bhm} \sum_{i=1}^{m} |\gamma_i| \; \| \hat{x}{}^{k+1}(\tau_{m-i}) \|. \qquad [6.36]$$

Inserting $e^{B(m-i)h} e^{-B(m-i)h} = 1$ into Eqn. (6.36) yields

$$e^{-Bhm} \sum_{i=1}^{m} |\gamma_i| \; e^{Bh(m-i)} e^{-Bh(m-i)} \| \hat{x}{}^{k+1}(\tau_{m-i}) \|. \qquad [6.37]$$

As

$$e^{-Bh(m-i)} \| \hat{x}{}^{k+1}(\tau_{m-i}) \| \leq \| \{\hat{x}{}^{k+1}(\tau_m)\} \|_B, \qquad [6.38]$$

the bound in Eqn. (6.37) can be simplified to

$$e^{-Bhm} \sum_{i=1}^{m} |\gamma_i| \; e^{Bh(m-i)} \| \{\hat{x}{}^{k+1}(\tau_m)\} \|_B. \qquad [6.39]$$

Reorganizing,

$$[\sum_{i=1}^{m} |\gamma_i| \, e^{-Bhi}] \, \| \{\hat{x}^{k+1}(\tau_m)\} \|_B. \qquad [6.40]$$

If $|\gamma_i|$ is bounded above by M, then the term in Eqn. (6.40) is bounded by

$$M[\sum_{i=1}^{m} e^{-Bhi}] \, \| \{\hat{x}^{k+1}(\tau_m)\} \|_B. \qquad [6.41]$$

Since e^{-Bhi} is always positive, the following inequality holds:

$$\sum_{i=1}^{m} e^{-Bhi} \leq \sum_{i=1}^{\infty} e^{-Bhi}, \qquad [6.42]$$

and from the infinite series summation formula

$$\sum_{i=1}^{\infty} e^{-Bhi} = \frac{e^{-Bh}}{1 - e^{-Bh}}.$$

Substituting the series summation formula into the bound in Eqn. (6.41) produces the conclusion of the lemma. ■

Proof of Theorem 6.2:

As before, only the Gauss-Seidel case will be proved. In order to insure charge conservation, the decomposed differential equations generated by the WR algorithm are solved using charge as the state variable. That is, the multistep integration algorithm is applied to

$$\dot{q}_i(x^{k,i}(t), u(t)) = f_i(x^{k,i}(t), u(t)) \qquad [6.43]$$

where $x^{k,i}(t)$, defined in Eqn. (6.25), is usually the vector of node voltages. A proof for the case where x is used as the state variable is given in [29]. Applying the multistep integration algorithm using the notation described above, and assuming $h_m = h$ for all m,

$$\rho(q_i(\hat{x}{}^{k,i}(\tau_m), u(\tau_m))) = h\sigma \left(f_i(\hat{x}{}^{k,i}(\tau_m), u(\tau_m))\right). \qquad [6.44]$$

Solving using the "inverse" operator yields

$$q_i(\hat{x}{}^{k,i}(\tau_m), u(\tau_m)) = h\rho^{-1}(\sigma(f_i(\hat{x}{}^{k,i}(\tau_m), u(\tau_m)))). \qquad [6.45]$$

Using the sum form for $\rho^{-1}\sigma$, and pulling out the leading term,

$$l_i(\hat{x}{}^{k,i}(\tau_m), u(\tau_m)) - h\gamma_0(f_i(\hat{x}{}^{k,i}(\tau_m), u(\tau_m))) - \qquad [6.46]$$

$$h\sum_{j=1}^{m}\gamma_j(f_i(\hat{x}{}^{k,i}(\tau_{m-j}), u(\tau_{m-j}))) + q_i(x(0), u(0)) = 0.$$

Define $F_i(\hat{x}(\tau_m))$ as

$$F_i(\hat{x}(\tau_m)) = q_i(\hat{x}(\tau_m), u(\tau_m)) - h\gamma_0 f_i(\hat{x}(\tau_m), u(\tau_m)) \qquad [6.47]$$

and define $b(\hat{x}_{past}, k) \in \mathbb{R}^n$ by

$$b_i(\hat{x}_{past}, k) = h\sum_{j=1}^{m}\gamma_j(f_i(\hat{x}{}^{k,i}(\tau_{m-j}), u(\tau_{m-j}))) + q_i(x(0), u(0)) \qquad [6.48]$$

where $(\hat{x}_{past}, k)$ is used to denote the fact that b is a function of $\hat{x}{}^k(\tau_l)$ and $\hat{x}{}^{k-1}(\tau_l)$ for all $l < m$. Then Eqn. (6.45) is like one iteration of the algebraic Gauss-Seidel algorithm applied to solving

$$F(\hat{x}(\tau_m)) - b(\hat{x}_{past}, k + 1) = 0 \qquad [6.49]$$

for $\hat{x}(\tau_m)$. Note that here, the right hand side moves because the past timepoints are moving with the iteration. As in Lemma 6.1, $\hat{x}^{k+1}$ can be written in terms of the map, $H(b(\hat{x}_{past}, k + 1))$, defined implicitly by the Gauss-Seidel relaxation algorithm applied to Eqn. (6.49),

$$\hat{x}^{k+1}(\tau_m) = H(b(\hat{x}_{past}, k + 1))\hat{x}^k(\tau_m). \qquad [6.50]$$

To prove that the iteration described by Eqn. (6.50) is a contraction mapping on the sequence $\{\hat{x}^k(\tau_m)\}$, it will be shown that given two arbitrary sequences, $\{\hat{x}^k(\tau_m)\}$ and $\{\hat{y}^k(\tau_m)\}$,

$$\max_m \, e^{-Bmh} \|\hat{x}^{k+1}(\tau_m) - \hat{y}^{k+1}(\tau_m)\| \le \qquad [6.51]$$

$$\gamma \, \max_m \, e^{-Bmh} \|\hat{x}^k(\tau_m) - \hat{y}^k(\tau_m)\|$$

where γ is some positive number strictly less than one. To simplify the equations, in the following we will use the notation

$$\max_m \, e^{-Bmh} \|\hat{x}^k(\tau_m)\| = \|\{\hat{x}^k(\tau_m)\}\|_B. \qquad [6.52]$$

To start, Eqn. (6.50) leads to the following equation for the difference between the two sequences at the mth step:

$$\hat{x}^{k+1}(\tau_m) - \hat{y}^{k+1}(\tau_m) = \qquad [6.53]$$

$$H(b(\hat{x}_{past}, k + 1))\hat{x}^k(\tau_m) - H(b(\hat{y}_{past}, k + 1))\hat{y}^k(\tau_m).$$

Breaking it into separate differences,

$$\hat{x}^{k+1}(\tau_m) - \hat{y}^{k+1}(\tau_m) = \tag{6.54}$$

$$H(b(\hat{x}_{past}, k+1))\hat{x}^k(\tau_m) - H(b(\hat{x}_{past}, k+1))\hat{y}^k(\tau_m)$$

$$+ H(b(\hat{x}_{past}, k+1))\hat{y}^k(\tau_m) - H(b(\hat{y}_{past}, k+1))\hat{y}^k(\tau_m)$$

and taking l_∞ norms,

$$\|\hat{x}^{k+1}(\tau_m) - \hat{y}^{k+1}(\tau_m)\|_\infty \le \tag{6.55}$$

$$\| H(b(\hat{x}_{past}, k+1))\hat{x}^k(\tau_m) - H(b(\hat{x}_{past}, k+1))\hat{y}^k(\tau_m) \|_\infty +$$

$$\| H(b(\hat{x}_{past}, k+1))\hat{y}^k(\tau_m) - H(b(\hat{y}_{past}, k+1))\hat{y}^k(\tau_m) \|_\infty.$$

At this point we will demonstrate that for small h, Eqn. (6.55) satisfies the assumptions of Lemma 6.1. It is assumed that the Jacobian of q with respect to x, $C(x(t), u(t))$, is strictly diagonally dominant uniformly in x. By definition, this assumption implies that there exists an $\varepsilon > 0$ such that

$$| C_{ii}(x(t), u(t)) | > \varepsilon + \sum_{j \ne i} | C_{ij}(x(t), u(t)) |. \tag{6.56}$$

Let $l > 0$ be the Lipschitz constant of f with respect to x. Assuming f is differentiable, if γ_0 of the multistep integration method is zero, and therefore the integration method is explicit, or if $h < |\frac{\varepsilon}{\gamma_0 l}|$, then the Jacobian of F defined in Eqn. (6.47), $\frac{\partial F}{\partial x}(\hat{x}(\tau_m))$, is strictly diagonally dominant.

Assuming h is small enough that $\frac{\partial F}{\partial x}(\hat{x}(\tau_m))$ is strictly diagonally dominant, then Lemma 6.1 can be applied to show

$$\| H(b(\hat{x}_{past}, k+1))\hat{x}^k(\tau_m) - H(b(\hat{x}_{past}, k+1))\hat{y}^k(\tau_m) \|_\infty \leq \qquad [6.57]$$

$$\gamma \| \hat{x}^k(\tau_m) - \hat{y}^k(\tau_m) \|_\infty$$

for some $\gamma < 1$ and

$$\| H(b(\hat{x}_{past}, k+1))\hat{y}^k(\tau_m) - H(b(\hat{y}_{past}, k+1))\hat{y}^k(\tau_m) \|_\infty \leq \qquad [6.58]$$

$$l_H \| b(\hat{x}_{past}, k+1) - b(\hat{y}_{past}, k+1) \|_\infty \| \hat{y}^k(\tau_m) \|_\infty$$

where l_H is the Lipschitz constant of H with respect to b.

Substituting Eqn. (6.57) and Eqn. (6.58) into Eqn. (6.55),

$$\| \hat{x}^{k+1}(\tau_m) - \hat{y}^{k+1}(\tau_m) \|_\infty \leq \qquad [6.59]$$

$$\gamma \| \hat{x}^k(\tau_m) - \hat{y}^k(\tau_m) \|_\infty + l_H M \| b(\hat{x}_{past}, k+1) - b(\hat{y}_{past}, k+1) \|_\infty$$

where $M = \max_m \| \hat{y}^k(\tau_m) \|_\infty$. Multiplying by e^{-Bmh} and taking the maximum over m,

$$\max_m e^{-Bmh} \| \hat{x}^{k+1}(\tau_m) - \hat{y}^{k+1}(\tau_m) \|_\infty \leq \qquad [6.60]$$

$$\gamma \max_m e^{-Bmh} \| \hat{x}^k(\tau_m) - \hat{y}^k(\tau_m) \|_\infty +$$

$$l_H M \max_m e^{-Bmh} \| b(\hat{x}_{past}, k+1) - b(\hat{y}_{past}, k+1) \|_\infty.$$

Or, using Definition 6.1,

$$\| \{ \hat{x}^{k+1}(\tau_m) - \hat{y}^{k+1}(\tau_m) \} \|_B \leq \qquad [6.61]$$

$$\gamma \, \| \{ \overset{\wedge k}{x}(\tau_m) - \overset{\wedge k}{y}(\tau_m) \} \|_B \; + \; l_H M \, \| \{ b(\hat{x}_{past}, k + 1) - b(\hat{y}_{past}, k + 1) \} \|_B.$$

The term in Eqn. (6.60) $b(\hat{x}_{past}, k + 1) - b(\hat{y}_{past}, k + 1)$ can be expanded using the definition of b in Eqn. (6.48) to

$$h \sum_{j=1}^{m} \gamma_j [f_i(\overset{\wedge k+1,i}{x}(\tau_{m-j}), u(\tau_{m-j})) - f_i(\overset{\wedge k+1,i}{y}(\tau_{m-j}), u(\tau_{m-j}))]. \qquad [6.62]$$

The β norm of the sequence whose terms are given in Eqn. (6.62) can be bounded using Lemma 6.2. That is,

$$\| \{ h \sum_{j=1}^{m} \gamma_j [f_i(\overset{\wedge k+1,i}{x}(\tau_{m-j}), u(\tau_{m-j})) - f_i(\overset{\wedge k+1,i}{y}(\tau_{m-j}), u(\tau_{m-j}))] \} \|_B \qquad [6.63]$$

$$\leq \; hN \frac{e^{-Bhm}}{1 - e^{-Bhm}} \| f_i(\overset{\wedge k+1,i}{x}(\tau_{m-j}), u(\tau_{m-j})) - f_i(\overset{\wedge k+1,i}{y}(\tau_{m-j}), u(\tau_{m-j})) \|_B,$$

where N is the $\max_m \gamma_m$. Using the triangle inequality and the Lipschitz property of f,

$$\| \{ h \sum_{j=1}^{m} \gamma_j [f_i(\overset{\wedge k+1,i}{x}(\tau_{m-j}), u(\tau_{m-j})) - f_i(\overset{\wedge k+1,i}{y}(\tau_{m-j}), u(\tau_{m-j}))] \} \|_B \; \leq \qquad [6.64]$$

$$hNl \frac{e^{-Bhm}}{1 - e^{-Bhm}} [\, \| \{ \overset{\wedge k+1}{x}(\tau_m) - \overset{\wedge k+1}{y}(\tau_m) \} \|_B + \| \{ \overset{\wedge k}{x}(\tau_m) - \overset{\wedge k}{y}(\tau_m) \} \|_B \,]$$

where l is the Lipschitz constant of f with respect to x. This bound can be used in Eqn. (6.61) to yield

$$(1 - hNl \frac{e^{-Bhm}}{1 - e^{-Bhm}}) \| \{ \overset{\wedge k+1}{x}(\tau_m) - \overset{\wedge k+1}{y}(\tau_m) \} \|_B \; \leq \qquad [6.65]$$

$$(\gamma - hNl\frac{e^{-Bhm}}{1 - e^{-Bhm}}) \| \{\hat{x}^k(\tau_m) - \hat{y}^k(\tau_m)\} \|_B.$$

We can reorganize Eqn. (6.65) into a more familiar form,

$$\| \{\hat{x}^{k+1}(\tau_m) - \hat{y}^{k+1}(\tau_m)\} \|_B \le \tag{6.66}$$

$$\frac{(\gamma + hNl\dfrac{e^{-Bhm}}{1 - e^{-Bhm}})}{(1 - hNl\dfrac{e^{-Bhm}}{1 - e^{-Bhm}})} \| \{\hat{x}^k(\tau_m) - \hat{y}^k(\tau_m)\} \|_B.$$

The factor γ is strictly less than one, so there exists an h_0 and a $b > 0$ such that

$$\frac{(\gamma + hNl\dfrac{e^{-bhm}}{1 - e^{-bhm}})}{(1 - hNl\dfrac{e^{-bhm}}{1 - e^{-bhm}})} < 1 \tag{6.67}$$

for all $h < h_0$, and this proves the theorem ∎.

SECTION 6.3 - THE MULTIRATE WR CONVERGENCE THEOREM

Theorem 6.1 suggests that the global-timestep discretized WR algorithm is not going to be any more efficient than the well-known relaxation-Newton algorithms described in Section 3.2, as the timestep constraints for the two methods are identical for the linear case. In fact, as Eqn. (6.10) indicates, WR is likely to be less efficient, because decomposition errors made in the first few timesteps propagate through the computations. The discretized WR algorithm has proved to be more efficient in practice for some types of problems because the discretized WR algorithm is intrinsically a multirate integration method. Because this is the key aspect of the WR algorithm, the rest of this chapter will be devoted to a proof that the discretized WR algorithm converges even when the timesteps for each subsystem are chosen independently.

Usually, choosing how to interpolate the discrete sequence of points produced by a numerical integration method is based only on what will produce attractive graphs of the computed solution. When multirate integration methods are applied to solving a system, interpolation plays a much more significant role. If two state variables in a system interact and they are computed using different timesteps, then to provide the value of one variable at the times required to compute the second variable, the first variable must interpolated. In the case of WR, if the interpolation is not done carefully, convergence may be destroyed.

In this section, a convergence theorem for systems in normal form will be presented that demonstrates the key role of interpolation in the convergence of the multirate discretized WR algorithm. The theorem guarantees that the discretized WR algorithm is a contraction mapping, assuming that the points produced by the numerical integration method are interpolated linearly. As the theorem proof will demonstrate, the linear interpolation has one particular property that aids convergence.

Consider the following system:

$$\dot{x}(t) = f(x(t), u(t)) \tag{6.68}$$

where $x(t) = (x_1(t),..., x_n(t))^T$; $x_i(t) \in \mathbb{R}^n$; $u(t) \in \mathbb{R}^m$ and is piecewise continuous; $f = (f_1(x),..., f_n(x))^T$; $f_i:\mathbb{R}^n \to \mathbb{R}$ is Lipschitz continuous. If the Gauss-Seidel WR algorithm is applied to Eqn. (6.68), the iteration equation for x_i is

$$\dot{x}_i^{k+1}(t) = f_i(x_1^{k+1}(t),..., x_i^{k+1}(t), x_{i+1}^k(t),..., x_n^k(t), u(t)). \tag{6.69}$$

If Eqn. (6.69) is solved numerically using a multistep integration algorithm with a fixed timestep h, the iteration equation becomes

$$p(\overset{\wedge k+1}{x_i}(\tau_m)) = h\sigma(f_i(\overset{\wedge k+1}{x_i}(\tau_m),..., \overset{\wedge k+1}{x_i}(\tau_m), \overset{\wedge k}{x_{i+1}}(\tau_m),..., \overset{\wedge k}{x_n}(\tau_m), u(\tau_m)). \quad [6.70]$$

If different timesteps are used to solve the differential equations associated with the x_j^k variables, $i \neq j$, then Eqn. (6.70) makes no sense, because τ_m for the jth equation may be different than τ_m for the ith equation. In order even to write down the equations for the multirate case, some kind of interpolation operator must first be defined.

Definition 6.2: Given a finite sequence $\{y(\tau_m)\}$ of M elements where $y(\tau_m) \in \mathbb{R}$ for all $m \leq M$, an interpolation function $I_t\{ \cdot \}$ on the sequence is any function that maps the sequence and the independent variable $t \in \mathbb{R}$, $t \in [\tau_0, \tau_M]$, into $\mathbb{R}$ such that $I_t\{ \cdot \}$ is continuous with respect to t, and $I_{\tau_j}\{y(\tau_j)\} = y(\tau_j)$.∎

As an example, the linear interpolation of a sequence at a given time $t \in [\tau_0, \tau_M]$ would be

$$I_t\{y(\tau_m)\} = y(\tau_j) + \frac{y(\tau_{j+1}) - y(\tau_j)}{(\tau_{j+1} - \tau_j)}(t - \tau_j) \quad [6.71]$$

where j is such that $\tau_j \leq t \leq \tau_{j+1}$.

In order to write a form of Eqn. (6.70) for the multirate case, we will denote τ_m for the ith equation as τ_m^i. Using this notation and the interpolation operator defined above, the index-filled iteration equation for $\overset{\wedge}{x_i^k}$ for the multirate fixed-timestep case is

$$p(\overset{\wedge k+1}{x_i}(\tau_m^i)) = h_i\sigma(f_i(I_{\tau_m^i}\{\overset{\wedge k+1}{x_1}(\tau_m^1)\},..., I_{\tau_m^i}\{\overset{\wedge k+1}{x_i}(\tau_m^i)\},$$

$$I_{\tau_m^i}\{\overset{\wedge k}{x_{i+1}}(\tau_m^{i+1})\},..., I_{\tau_m^i}\{\overset{\wedge k}{x_n}(\tau_m^n)\}, u(\tau_m^i))$$

where h_i is the fixed timestep for the ith system. Using the inverse operator as in Eqn. (6.18),

$$\hat{x}_i^{k+1}(\tau_m^i) = h_i \rho^{-1}(\sigma(f_i(I_{\tau_m^i}\{\hat{x}_1^{k+1}(\tau_m^1)\}, ..., I_{\tau_m^i}\{\hat{x}_i^{k+1}(\tau_m^i)\}, \qquad [6.72]$$

$$I_{\tau_m^i}\{\hat{x}_{i+1}^{k}(\tau_m^{i+1})\}, ..., I_{\tau_m^i}\{\hat{x}_n^{k}(\tau_m^n)\}, u(\tau_m^i))))$$

The proof of Theorem 6.2 demonstrated that the fixed global-timestep discretized WR algorithm is a contraction mapping in an $l_\infty \beta$ norm on the sequence (see Definition 6.1). In the multirate case, this is not sufficient. Since interpolated as well as sequence values are used by subsystems, a convergence proof must take into account the effect of the the interpolation on the sequence. The approach that will be used in the proof that follows is to view the multirate discretized WR algorithm, which necessarily includes an interpolation operator, as a map of continuous functions on $[0,T]$ to continuous functions. The implicitly defined map can be derived by applying the interpolation operator to both sides of Eqn 6.72 to yield

$$I_t\{\hat{x}_i^{k+1}(\tau_m^i)\} = I_t\{h_i \rho^{-1}(\sigma(f_i(I_{\tau_m^i}\{\hat{x}_1^{k+1}(\tau_m^1)\}, ..., I_{\tau_m^i}\{\hat{x}_i^{k+1}(\tau_m^i)\}, \qquad [6.73]$$

$$I_{\tau_m^i}\{\hat{x}_{i+1}^{k}(\tau_m^{i+1})\}, ..., I_{\tau_m^i}\{\hat{x}_n^{k}(\tau_m^n)\}, u(\tau_m^i))))\}.$$

To prove the convergence of the relaxation, the usual continuous-time $l_\infty \beta$ norm can be used,

$$\|x\|_B = \max_{[0,T]} e^{-Bt}[\max_i | I_t\{x_i(\tau_m^i)\} |], \qquad [6.74a]$$

or equivalently,

$$\|x\|_B = \max_i [\ \max_{[0,T]} e^{-Bt} |\ I_t\{x_i(\tau_m^i)\}\ |\ ],$$ [6.74b]

where x is used to denote the vector function on $[0,T]$ defined by $x(t) = (I_t\{x_1(\tau_m^i)\},..., I_t\{x_n(\tau_m^i)\})^T$.

Under certain conditions Eqn. (6.73) is a contraction map in the β norm of Eqn. (6.74). To prove this, Eqn. (6.73) will be applied to two sequences $\{x^k(\tau_m)\}$ and $\{y^k(\tau_m)\}$. The difference between Eqn. (6.73) applied to the two sequences is

$$I_t\{\hat{x}_i^{k+1}(\tau_m^i)\} - I_t\{\hat{y}_i^{k+1}(\tau_m^i)\} =$$ [6.75]

$$I_t\{h_i\rho^{-1}(\sigma(f_i(\ I_{\tau_m^i}\{\hat{x}_1^{k+1}(\tau_m^1)\},..., I_{\tau_m^i}\{\hat{x}_i^{k+1}(\tau_m^i)\},$$

$$I_{\tau_m^i}\{\hat{x}_{i+1}^{k}(\tau_m^{i+1})\},..., I_{\tau_m^i}\{\hat{x}_n^{k}(\tau_m^n)\}, u(\tau_m^i)\ )))\}$$

$$-$$

$$I_t\{h_i\rho^{-1}(\sigma(f_i(\ I_{\tau_m^i}\{\hat{y}_1^{k+1}(\tau_m^1)\},..., I_{\tau_m^i}\{\hat{y}_i^{k+1}(\tau_m^i)\},$$

$$I_{\tau_m^i}\{\hat{y}_{i+1}^{k}(\tau_m^{i+1})\},..., I_{\tau_m^i}\{\hat{y}_n^{k}(\tau_m^n)\}, u(\tau_m^i)\ )))\}.$$

It is possible to simplify Eqn. (6.75) by only considering interpolation operators that are linear functions of the sequence. By this we do not mean to limit consideration only to linear interpolation, but to those interpolation functions for which

$$I_t\{x(\tau_m)\} - I_t\{y(\tau_m)\} = I_t\{x(\tau_m) - y(\tau_m)\}$$

and

$$I_t\{\alpha x(\tau_m)\} = \alpha I_t\{x(\tau_m)\}$$

where $\{x(\tau_m)\}$, $\{y(\tau_m)\}$ are sequences in $\mathbb{R}$, and $\alpha \in \mathbb{R}$. For example, any of the spline[58] or polynomial interpolation operators are linear functions of the sequence. Exploiting this linearity in Eqn. (6.75) leads to

$$I_t\{\hat{x}{}_i^{k+1}(\tau_m^i) - \hat{y}{}_i^{k+1}(\tau_m^i)\} = \tag{6.76}$$

$$h_i I_t\{ \ \rho^{-1}(\sigma([$$

$$f_i(\ I_{\tau_m^i}\{\hat{x}{}_1^{k+1}(\tau_m^1)\},...,\ I_{\tau_m^i}\{\hat{x}{}_i^{k+1}(\tau_m^i)\},\ I_{\tau_m^i}\{\hat{x}{}_{i+1}^{k}(\tau_m^{i+1})\},...,\ I_{\tau_m^i}\{\hat{x}{}_n^{k}(\tau_m^n)\},\ u(\tau_m^i)\ )\ -$$

$$f_i(\ I_{\tau_m^i}\{\hat{y}{}_1^{k+1}(\tau_m^1)\},...,\ I_{\tau_m^i}\{\hat{y}{}_i^{k+1}(\tau_m^i)\},\ I_{\tau_m^i}\{\hat{y}{}_{i+1}^{k}(\tau_m^{i+1})\},...,\ I_{\tau_m^i}\{\hat{y}{}_n^{k}(\tau_m^n)\},\ u(\tau_m^i)\ )\ ]))\ \}.$$

We have been able to prove that the multirate discretized WR algorithm is a contraction mapping in the β norm of Eqn. (6.74) only under the strong assumption that the interpolation operator is the linear interpolation operator as in Eqn. (6.71). Although practical experience indicates that other interpolation operators can be used successfully, we have not succeeded in generalizing the following theorem to include those operators.

Theorem 6.3: If the interpolation in Eqn. (6.73) is linear interpolation, then there exists a collection of timesteps $h_{i0} > 0$, $i \in \{1,...,n\}$, such that if $0 < h_i \le h_{i0}$ for all i, then the multirate fixed-timestep discretized WR algorithm converges with respect to the interpolated sequences. ∎

The following simple lemmas will be useful in the proof.

Lemma 6.3: If $I_t\{\ \bullet\ \}$ is the linear interpolation operator (as in Eqn. (6.73)), then given two sequences $\{x(\tau_m)\}$ and $\{y(\tau_m)\}$, if $x(\tau_i) \ge y(\tau_i)$ for all i then

$I_t\{x(\tau_m)\} \geq I_t\{y(\tau_m)\}$ for all t for which the interpolation is defined. In addition, if $x(\tau_m) = K$, $K \in \mathbb{R}$, for all $m \leq m'$, then $I_t\{x(\tau_m)\} = K$ for all $t \in [0, \tau_{m'}]$. ∎

Lemma 6.3 follows directly from the definition of linear interpolation. As will be shown in the proof, this is the key property of linear interpolation with respect Theorem 6.3.

Lemma 6.4: If $I_t\{ \cdot \}$ is the linear interpolation operator and $\{x(\tau_m)\}$ is a sequence in $\mathbb{R}$, then

$$\max_{[0,T]} e^{-Bt} | I_t\{ \sum_{l=0}^{m_i} |\gamma_l| \; |I_{(m-l)h}\{x(\tau_m)\}| \; \} | \leq$$

$$\frac{1}{1 - e^{-Bh}} M \max_{[0,T]} e^{-Bt} | I_t\{ |x(\tau_m)| \} |$$

where $M = \max_l |\gamma_l|$. ∎

The proof of Lemma 6.4 parallels the arguments given in the proof of Lemma 6.2, and is omitted.

Proof of Theorem 6.3:

Expanding the $\rho^{-1}\sigma$ operator in Eqn. (6.76) into its sum form leads to

$$I_t\{\hat{x}_i^{k+1}(\tau_m^i) - \hat{y}_i^{k+1}(\tau_m^i)\} =$$

$$h_i I_t\{ \sum_{l=0}^{m_i} \gamma_l (f_i(I_{\tau_{m-l}^i}\{\hat{x}_1^{k+1}(\tau_m^1)\},..., I_{\tau_{m-l}^i}\{\hat{x}_i^{k+1}(\tau_m^i)\},$$

$$I_{\tau_{m-l}^i}\{\hat{x}_{i+1}^k(\tau_m^{i+1})\},..., I_{\tau_{m-l}^i}\{\hat{x}_n^k(\tau_m^n)\}, u(\tau_m^i))$$

$$- (f_i(I_{\tau^i_{m-1}}\{\hat{y}_1^{k+1}(\tau^1_m)\},..., I_{\tau^i_{m-1}}\{\hat{y}_i^{k+1}(\tau^i_m)\},$$

$$I_{\tau^i_{m-1}}\{\hat{y}_{i+1}^{k}(\tau^{i+1}_m)\},..., I_{\tau^i_{m-1}}\{\hat{y}_n^{k}(\tau^n_m)\}, u(\tau^i_m)) \ ] \ \}.$$

Using Lemma 6.3 and the Lipschitz continuity property of f,

$$|I_t\{\hat{x}_i^{k+1}(\tau^i_m) - \hat{y}_i^{k+1}(\tau^i_m)\}| \leq \tag{6.77}$$

$$|I_t\{ \sum_{l=0}^{m_i} |\gamma_l| \ [\sum_{j=1}^{i} h_i L_{ij} \ |I_{\tau^i_{m-1}}\{\hat{x}_j^{k+1} - \hat{y}_j^{k+1}\}| \ + $$

$$\sum_{j=i+1}^{n} h_i L_{ij} \ |I_{\tau^i_{m-1}}\{\hat{x}_j^{k} - \hat{y}_j^{k}\}| \ ] \ \}|$$

where L_{ij} is the Lipschitz constant of f_i with respect to x_j. Reorganizing, and exploiting the general linearity property of the interpolation operator and the triangle inequality,

$$|I_t\{\hat{x}_i^{k+1}(\tau^i_m) - \hat{y}_i^{k+1}(\tau^i_m)\}| \leq \tag{6.78}$$

$$\sum_{j=1}^{i} h_i L_{ij} |I_t\{ \sum_{l=0}^{m_i} |\gamma_l| \ |I_{\tau^i_{m-1}}\{\hat{x}_j^{k+1} - \hat{y}_j^{k+1}\}| \ \}| \ + $$

$$\sum_{j=i+1}^{n} h_i L_{ij} |I_t\{ \sum_{l=0}^{m_i} |\gamma_l| \ |I_{\tau^i_{m-1}}\{\hat{x}_j^{k} - \hat{y}_j^{k}\}| \ \}|.$$

Multiplying by e^{-Bt} and taking maximums,

$$\max_{[0,T]} e^{-Bt} |I_t\{\hat{x}_i^{k+1}(\tau^i_m) - \hat{y}_i^{k+1}(\tau^i_m)\}| \leq \tag{6.79}$$

$$\sum_{j=1}^{i} h_i L_{ij} \max_{[0,T]} e^{-Bt} | I_t \{ \sum_{l=0}^{m_i} |\gamma_l| \; | I_{\tau_{m-l}^i} \{ \hat{x}_j^{k+1} - \hat{y}_j^{k+1} \} | \; \} | \; + $$

$$\sum_{j=i+1}^{n} h_i L_{ij} \max_{[0,T]} e^{-Bt} | I_t \{ \sum_{l=0}^{m_i} |\gamma_l| \; | I_{\tau_{m-l}^i} \{ \hat{x}_j^{k} - \hat{y}_j^{k} \} | \; \} |. $$

Applying Lemma 6.4,

$$\max_{[0,T]} e^{-Bt} | I_t \{ \hat{x}_i^{k+1}(\tau_m^i) - \hat{y}_i^{k+1}(\tau_m^i) \} | \; \le \qquad [6.80]$$

$$\left[\frac{Mh_i}{1 - e^{-Bh_i}} \sum_{j=1}^{i} L_{ij} \right] \| \hat{x}^{k+1} - \hat{y}^{k+1} \|_B \; + \; \left[\frac{Mh_i}{1 - e^{-Bh_i}} \sum_{j=i+1}^{n} L_{ij} \right] \| \hat{x}^{k} - \hat{y}^{k} \|_B$$

where $\| \hat{x}^k - \hat{y}^k \|_B$ is the $l_\infty \beta$ norm defined in Eqn. (6.74b).

For any $\delta > 0$ there exists a collection of steps $\{h_{1_0}, ..., h_{n_0}\}$, all strictly positive, and a $B > 0$ such that

$$\delta > \frac{Mh_i \sum_{j=1}^{i} L_{ij}}{1 - e^{-Bh_i}} \qquad [6.81]$$

for all $h_i \le h_{i_0}$, for all i. Substituting into Eqn 6.80,

$$\max_{[0,T]} e^{-Bt} | I_t \{ \hat{x}_i^{k+1}(\tau_m^i) - \hat{y}_i^{k+1}(\tau_m^i) \} | \; \le \qquad [6.82]$$

$$\delta \| \hat{x}^{k+1} - \hat{y}^{k+1} \|_B \; + \; \delta \| \hat{x}^{k} - \hat{y}^{k} \|_B.$$

Since Eqn. (6.81) holds for all i,

$$\|\hat{x}{}^{k+1} - \hat{y}{}^{k+1}\|_B \leq \delta \|\hat{x}{}^{k+1} - \hat{y}{}^{k+1}\|_B + \delta \|\hat{x}{}^{k} - \hat{y}{}^{k}\|_B. \quad [6.83]$$

Reorganizing,

$$\|\hat{x}{}^{k+1} - \hat{y}{}^{k+1}\|_B \leq \frac{\delta}{1-\delta} \|\hat{x}{}^{k} - \hat{y}{}^{k}\|_B. \quad [6.84]$$

Clearly, there exists a $\delta > 0$ such that Eqn. (6.84) is a contraction $(\frac{\delta}{1-\delta} < 1)$. Let $\overset{\wedge}{\delta}$ be that δ. Since there exists a $B > 0$ and a collection of h_{i_0}'s > 0 such that Eqn. (6.84) holds for $\delta = \overset{\wedge}{\delta}$ for all $0 < h_i \leq h_{i_0}$, the theorem is proved. ■

Perhaps the most surprising aspect of the proof of Theorem 6.3 is that the ratio of the timesteps from one system to the next does *not* seem to play a role. This is an important observation given that the discretized WR algorithm was developed to allow different subsystems to take vastly different timesteps. If a large ratio between timesteps destroyed the WR convergence, then the applicability of the WR algorithm to multirate problems would be limited.

A second important observation is that the only property of linear interpolation used in the course of the proof was that stated in Lemma 6.3. Therefore, other interpolations that have this property will work as well. Higher-order polynomial interpolation functions do not have the property stated in Lemma 6.3, but as they are substantially more accurate than linear interpolation, they are extremely useful. An extension of the above theorem to general polynomial interpolation does not seem to be straightforward, and may call for an entirely different approach.

CHAPTER 7 - THE IMPLEMENTATION OF WR

In this chapter, a description of the implementation of the WR algorithm in the RELAX2 program is given. We start with a brief overview of the steps performed in the RELAX2 program when simulating a circuit. A detailed description of the major steps is contained in the sections that follow.

The first step in simulating a circuit using the RELAX2 program is to create the circuit description file. In this file a user must specify device model parameters, circuit topology, analysis specifications, and plotting requests. The circuit topology can be described in as hierarchical or as flat a form as the user desires[60]. This circuit description file is used as an input to the RELAX2 program, whose first step is to flatten the hierarchy.

Before applying the WR algorithm, the flattened circuit is decomposed into a collection of *subcircuits*. This is done by partitioning the circuit into clusters of tightly coupled nodes. Then the elements (e.g. transistors, resistors, capacitors) that connect to any of the nodes in a given cluster are gathered together to make the subcircuits. Once the entire circuit has been carved up into subcircuits, the subcircuits are ordered, or scheduled, starting with subcircuits that are connected to the user-defined inputs and then following the natural directionality of the circuit as much as possible.

After a large circuit has been broken up into subcircuits, and these subcircuits have been ordered, the RELAX2 program begins the waveform relaxation process. An initial guess is made for each of the node voltage waveforms. Then the numerical solution for each of the subcircuits is computed in the order determined above. The computation is performed using a variable-timestep trapezoidal rule numerical integration algorithm, with local truncation error timestep control[1]. To perform the numerical integration, those nodes in the subcircuit that were not part of the cluster around which the subcircuit was built are treated as external time-varying voltage sources. The values for the external voltage sources are either the initial guess waveforms, or, if the subcircuit containing the external node was simulated previously, that computed waveform. As the node waveforms are computed, they replace the existing waveforms (initial guesses or previous iterations), and the process is repeated until the waveforms converge.

As mentioned in Chapter 5, the WR algorithm becomes inefficient when used to simulate digital circuits with logical feedback(e.g. finite-state machines, ring oscillators, etc.) for many cycles, because the relaxation converges in a very nonuniform manner. For this reason, the RELAX2 program does not actually perform the relaxation iterations by computing the transient behavior of each subcircuit for the entire user-defined simulation interval. Instead, the RELAX2 program uses a modified WR algorithm[17], in which the relaxation is only performed for a small piece of the user-defined simulation interval at a time. Exactly how large a piece of the waveform, referred to as a *window*, to use is determined automatically at the beginning of every WR iteration.

If the WR algorithm is applied to very large circuits, it is often the case that some pieces of the circuit will converge much more rapidly than others. This phenomenon, called partial waveform convergence, can be exploited to improve the overall efficiency of the WR method. The details of the algorithm for avoiding recomputing the waveforms that have already converged are given in Section 7.5.

As a final point, in Chapter 6 it was mentioned that when the WR iteration equations are solved using a numerical integration algorithm, the resulting discretized WR algorithm is not guaranteed to converge *unless* the discretization error is driven to zero with the iterations. To insure that the iterations do converge, the RELAX2 program reduces the acceptable local truncation error criteria used for selecting the numerical integration timesteps as the iterations in a given window progress.

SECTION 7.1 - PARTITIONING MOS CIRCUITS

As was shown in Section 5.2, the convergence of WR is greatly accelerated if groups of tightly coupled nodes are solved together as one subsystem or subcircuit. For this reason, the RELAX2 program groups together tightly coupled nodes into subcircuits before beginning the relaxation process. The algorithms used to partition large MOS circuits in the RELAX2 program are based on Algorithm 5.1 for partitioning linear algebraic systems, and on Theorems 5.3 and 5.4 which relate the problem of partitioning linear algebraic systems to partitioning linear differential systems.

MOS circuits are not linear, so the ideas presented in Section 5.2 must be modified if they are to be applied to nonlinear systems. The RELAX2 program uses several conservative heuristics (conservative in the sense that they tend to err on the side of producing subcircuits which are larger than needed) to handle the nonlinear MOS transistors. The first heuristic is that each of the MOS transistors is initially treated as a nonlinear resistor between the transistor's source and drain, and the couplings between the gate and source and between the gate and drain are considered separately, during scheduling (See Section 7.2). With this simplification, Algorithm 7.1 below, an algorithm for partitioning circuits with two-terminal linear and nonlinear resistances, is applied.

The Norton equivalent conductance at a node, *Geq*, used in Algorithm 7.1 can be computed using a simple recursive formula if the network is a series-parallel network[67](Algorithm 7.2). Note that this recursion will not be very

Algorithm 7.1 - (Conductance Partitioning)
    for each (*conductive element* in *the circuit*) {
        $node_1$ ← one of the element terminal nodes.
        $node_2$ ← the other element terminal node.
        g_3 ← maximum element conductance (over all v).
        Remove the element from the circuit.
        Replace the remaining circuit conductances with their minimum values.
        g_1 ← the Norton equivalent conductance at $node_1$.
        g_2 ← the Norton equivalent conductance at $node_2$.

        If ($\dfrac{g_3}{(g_2 + g_3)} \dfrac{g_3}{(g_1 + g_3)}$ > α) Tie $node_1$ to $node_2$.

                α is a desired WR iteration factor, typically 0.3

    }
    ∎

deep. The recursion will stop at any terminal of an MOS transistor, because the
minimum conductance of the MOS transistor is zero.

If the circuit is not series-parallel, Algorithm 7.2 can still be used if the
recursion is truncated in such a way that no circuit node is visited twice. In this
case, only an estimate of the Norton equivalent will be computed.

Algorithm 7.2 - (Norton Equivalent Conductance for Node i)
    Geq ← 0.0
    foreach (*conductive element* incident *at node i*) {
        G ← element conductance
        $nodej$ ← the conductive element's other node.
        If ($nodej$ is a voltage source node) {
            Geq → $Geq + G$
        }
        else {
            $Geqj$ ← Norton equivalent conductance at $nodej$ with this element
            removed. Geq ← $Geq + (G \times Geqj)/(G + Geqj)$
        }
    }
    ∎

The conductance partitioning algorithm is justified by Theorem 5.3, that
the WR iteration factor is bounded below by the iteration factor for solving just

the algebraic portion of the problem. Theorem 5.4 suggests that an analogous algorithm to Algorithm 7.1 be constructed for the capacitive elements in the circuit. Since the capacitance problem is almost identical in nature to the conductance problem, a capacitance partitioning algorithm can follow almost the same strategy as the conductance partitioning algorithm. The difference is that instead of comparing floating capacitances to Norton equivalent conductances, floating capacitances are compared to Norton equivalent capacitances. These equivalent capacitances are entirely analogous to the equivalent conductances, and can be computed using the same recursive approach as in Algorithm 7.2.

The RELAX2 program uses both conductance and capacitive partitioning, and forms subcircuits from the union of the two results. The algorithm has been applied to a wide variety of MOS digital circuits, including a large VHSIC memory circuit with 2900 nodes, 600 MOS devices, and over 3500 parasitic components. The results have always matched the best attempts at hand partitioning, in as many instances as we had the patience to check. However, as the method is applied to large variety of problems, there may be cases where subcircuits produced become quite large. Should this be the case, the present simple algorithm could be extended, so that an additional pass is made over the excessively large subcircuits, to subpartition them using more sophisticated algorithms. In particular, the subpartitioning could use better estimates of the equivalent conductances and capacitances, as the present algorithm may be unnecessarily conservative.

SECTION 7.2 - ORDERING THE SUBSYSTEM COMPUTATION

As described in Section 5.2, resistors and capacitors do not exhibit the kind of unidirectional coupling that is of concern to an ordering algorithm which is attempting to order the system to be as block lower triangular as possible. In fact, the only element type of concern to the ordering algorithm is the transistor, because it exhibits unidirectional coupling. That is, the drain and source terminals of an MOS transistor are strongly dependent on the gate terminal of the transistor, but the gate is almost independent of the drain and source. Clearly, this implies that the subcircuits containing the given transistor's drain or source should be analyzed after the subcircuit containing the given transistor's gate.

To devise an algorithm to carry out this task, it is convenient to introduce a *dependency graph* of the partitioned circuit. If we represent the circuit with a directed graph $G(X, E)$, where the set of nodes of the graph X is in one-to-one correspondence with the subcircuits obtained by a partitioner, and where there is a directed edge between the node corresponding to subcircuit i and the node corresponding to subcircuit j if there is a transistor whose gate is in subcircuit i and whose drain or source is in subcircuit j. If the graph is acyclic, it can be levelized; i.e. all the nodes can be ordered in *levels* so that a node in level i can have incoming edges only from nodes in levels lower than i. The ordering so obtained is the one used by RELAX2 to process the subcircuits.

However, there may be cases where cycles exist in the graph. In this case, either the subcircuit definitions are changed by grouping two or more subcircuits together, effectively performing part of the partitioning task(as eluded to in Section 7.1), or edges of the graph are discarded to remove the cycles. In either case, at the end of this process an acyclic graph and an ordering of the subcircuits corresponding to the leveling of the (perhaps altered) graph is obtained.

One decision remains: when to repartition to remove a feedback loop rather than breaking the loop. As the example of Section 5.1 indicates, if signal propagation around the feedback loop is fast compared to the size of the window, the relaxation convergence will be slow and nonuniform.. For this reason, the ordering algorithm makes the decision about partitioning based on an estimate of the delay around the feedback loop. If the loop delay is smaller than one percent(somewhat arbitrarily chosen) of the simulation interval, the feedback loop is removed by repartitioning. If the delay is longer, then the feedback loop is broken by removing an edge from the directed graph.

Algorithm 7.3 - (RELAX2 Subcircuit Ordering Algorithm)
Initialization.
ordered__list = NULL;
unordered__list = List of subcircuits from the partitioner;
Main Loop.
while (unordered__list ≠ NULL) {
  none__ordered = FALSE;
  while (none__ordered = FALSE) {
    none__ordered = TRUE;
    for each (subcircuit in the unordered__list) {
      if (all subcircuits on incoming arcs are on ordered__list) {
        none__ordered = FALSE;
        append__to__end__of__ordered__list(subcircuit);
        delete__from__unordered__list(subcircuit);
      }
    }
  }
  if (unorder__list ≠ NULL) { *Must be a feedback loop.*
    found__loop = FALSE;
    depth = 1;
    while (found__loop == FALSE) {
      depth = depth + 1;
      for each (subcircuit in the unordered list) {
        if (there exists a loop of length = depth) {
          found__loop = TRUE;
          if (delay around the loop > 0.01 * the simulation interval) {
            break the loop
          }
          else {
            collapse loop into one subcircuit.
          }
        }
      }
    }
  }
}
∎

SECTION 7.3 - COMPUTATION OF THE SUBSYSTEM WAVEFORMS

As in standard circuit simulators, the RELAX2 program solves Eqn. (4.4) using a numerical integration method with varying timesteps. Since the major aim of the RELAX2 program is to simulate digital MOS circuits, the integration method was chosen based on how effectively it solves problems with the properties of digital MOS circuits. Digital circuits are very stiff; therefore only A-stable integration methods were considered. In addition, digital circuits contain very rapid transitions, and low-order one-step integration methods are usually suggested for such problems. Although the implicit-Euler method is computationally the simplest A-stable one-step method, the trapezoidal rule, an A-stable second-order one-step method, was chosen instead because of its better accuracy.

There is a second important reason for chosing the trapezoidal integration algorithm over the implicit-Euler formula. If the WR algorithm is used to solve the system, and a numerical integration method is used to solve the WR iteration equations, then the iteration factor (Definition (5.3)) for the relaxation will be a function of the integration method. For circuit simulation problems, the iteration factor for the trapezoidal rule will usually be smaller than for the implicit-Euler method, and a smaller iteration factor implies faster convergence.

To show this, consider the WR algorithm applied to an equation system of the form $\dot{v}(t) = Av(t)$ where $A \in \mathbb{R}^{n \times n}$ is such that $A := L + D + U$ where L, D, and U are the strictly lower-triangular, diagonal, and strictly upper-triangular portions of A respectively, and assume that L and U are positive matrices and D is a negative matrix. Such a problem would be generated by a circuit made up of grounded linear capacitors and linear two-terminal resistors. The Gauss-Jacobi WR iteration equation for this example is

$$\dot{v}^{k+1} = Dv^{k+1} + (L + U)v^k. \qquad [7.1]$$

Now consider the first timestep of the implicit-Euler discretized Gauss-Jacobi WR algorithm:

$$v^{k+1}(h) - v_0 = h[Dv^{k+1}(h) + (L + U)v^k(h)] \qquad [7.2]$$

which can be reorganized as

$$v^{k+1} = [I - hD]^{-1}[h(L + U)v^k + v_0].$$

The iteration factor for the relaxation is the norm of the matrix $[I - hD]^{-1}h(U + L)$.

Applying the trapezoidal rule yields

$$v^{k+1}(h) - v_0 = 0.5h\,Dv^{k+1}(h) + 0.5h\,(L + U)v^k(h) + 0.5h\,Av_0 \qquad [7.3]$$

which can be reorganized as

$$v^{k+1} = [I - 0.5hD]^{-1}[0.5h(L + U)v^k + (I + 0.5hAv_0].$$

The iteration factor for the relaxation is the norm of $[I - 0.5hD]^{-1}0.5h(L + U)$. Since D is a negative matrix and L and U are positive matrices, the following inequality holds for any induced norm:

$$\| [I - 0.5hD]^{-1}0.5h(L + U) \| < \| [I - hD]^{-1}h(U + L) \|$$

and therefore the iteration factor for the trapezoidal-rule discretized WR algorithm is smaller than the iteration factor for the implicit-Euler discretized WR algorithm.

Given a timestep h, the trapezoidal integration method applied to Eqn. (3.2) yields

$$q(t + h) - q(t) - 0.5h(\hat{f}(q(t + h),u) + \hat{f}(q(t), u)) = 0. \qquad [7.4]$$

The above equation is a nonlinear algebraic equation in q. The user is usually more interested in the voltage, so before solving Eqn. (7.4) we substitute for q in terms of v:

$$q(v(t + h)) - q(v(t)) - 0.5h[\, f(v(t + h),u) + f(v(t), u)\,] = 0. \qquad [7.5]$$

In Eqn. (7.5) $v(t)$ and $q(t)$ are known, and the equation must be solved to compute $v(t + h)$. Nonlinear algebraic systems generated by integration methods are usually solved using the iterative Newton-Raphson method. This is because Newton methods have quadratic convergence properties and because they are guaranteed to converge if the initial guess is close enough to the correct solution. The general Newton-Raphson iteration equation to solve $F(x) = 0$ is

$$J_F(x^l)\,(x^l - x^{l-1}) = -F(x^{l-1}) \qquad [7.6]$$

where J_F is the Jacobian of F with respect to x. The iteration is continued until $\| x^l - x^{l-1} \| < \varepsilon$ and $F(x)$ is close enough to 0. If the Newton algorithm is used to solve Eqn. (7.5) for $v(t + h)$, the residue, $F(v^l(t + h))$, is

$$F(v^l(t + h)) = q(v^l(t + h)) - q(v(t)) - 0.5h(\, f(v^l(t + h),u) + f(v(t), u)\,) \qquad [7.7]$$

and the Jacobian of $F(v^l(t + h))$, $J_F(v^l(t + h))$ is

$$J_F(v^l(t + h)) = C(v^l(t + h),u) + 0.5h \frac{\partial f(v^l(t + h), u)}{\partial v}.$$

Then $v^{l+1}(t + h)$ is derived from $v^l(t + h)$ by solving the linear system of equations

$$J_F(v^l(t + h))\,[v^{l+1}(t + h) - v^l(t + h)] = -F(v^l(t + h)) \qquad [7.8]$$

The Newton iteration is continued until $\| v^{l+1}(t + h) - v^l(t + h) \| < \varepsilon$ and $F(v^l(t + h))$ is close enough to zero.

Each iteration of the Newton algorithm requires a function evaluation, a Jacobian evaluation, and a matrix solution. For the algebraic systems generated by the numerical integration of MOS digital circuits it is often inefficient to evaluate the Jacobian every Newton iteration. If the Jacobian is re-evaluated only every few Newton iterations[27], the number of iterations required to achieve convergence is usually unchanged and the computation required is significantly reduced. Not only are Jacobian evaluations skipped, but if the matrix solution is computed by LU factorization[44], subsequent matrix solutions using the same matrix can skip the LU factorization step. By default, the RELAX2 program evaluates the Jacobian every third Newton iteration; this choice is based on experimental evidence such as the several examples given in Table 7.1. However, the efficiency of such an approach is problem dependent, so the number of Newton iterations per Jacobian evaluation can be changed by the user.

TABLE 7.1 - CPU TIME VS # ITERS/JACOBIAN EVALUATION					
Circuit	Devices	1	2	3	4
Ring Osc.	7	0.95s	0.77s	0.71s	0.75s
Oper. Amp	25	6.28s	5.2s	4.52s	4.67s
flip-flop	33	20.47s	16.82s	13.93s	13.67s
Cmos Memory	621	1080s	976s	885s	886s

On Vax11/780 running Unix

The integration method used in the SPICE2 program is very similar to the direct method used in RELAX2. Both use the trapezoidal integration formula with some form of local-truncation-error-based timestep control algorithm, the Newton method to solve the algebraic system, and sparse LU factorization to perform the matrix solution. However, as can be seen from Table 7.2, the RELAX2 program, using the direct method described above, is eight to twenty times faster than the SPICE2 program. This can be attributed to many factors. The first is that RELAX2 is written in "C" while SPICE2 is in FORTRAN, and "C" programs under the UNIX operating system run almost a factor of two times faster than FORTRAN programs. The other factor of four to ten is due to more sophisticated programming techniques, more efficient equation formulation, the modified Newton method mentioned above, and better numerical integration error control.

TABLE 7.2 - RELAX2 (DIRECT) VS SPICE				
Circuit	Devices	SPICE2	RELAX2	Ratio
Ring Osc.	7	17s	0.75s	22
Op-amp	25	42s	5s	8
uP Control	232	1400s	90s	15
Cmos Memory	621	10400s	995s	10
4-bit Counter	259	4300s	540s	8
Encode-Decode	1326	115,840s	5000s	23

On Vax11/780 running Unix

It should be pointed out that without using a fundamentally new circuit simulation method, but just by carefully exploiting some very general properties of MOS digital circuits, almost an order of magnitude decrease in computation time has been achieved over the more general SPICE2 program.

SECTION 7.4 - WINDOW SIZE DETERMINATION

As mentioned in Section 5.2, the WR algorithm used in RELAX2 becomes inefficient when used to simulate digital circuits with logical feedback(e.g. finite-state machines, ring oscillators, etc.) for many cycles. However, the WR algorithm can still be very efficient if the relaxation is only performed on a piece of the waveform to be computed at a time. For general circuits, an ideal situation would be to break the simulation interval into windows over which every timepoint of the iteration waveform moves closer to the correct solution. However, if the windows are too small some of the advantages of waveform relaxation are lost. One cannot take advantage of a digital circuit's natural latency over the entire waveform, but only in that window; the scheduling overhead increases when the windows become smaller, as each circuit lump must be scheduled once for each window, and if the windows are made very small, timesteps chosen to calculate the waveforms will be limited by the window size rather than by the discretization error, and unnecessary calculations will be performed.

Rather than use a conservative *a priori* lower bound as given in Theorem 5.2, in the RELAX2 program the window size is determined dynamically by two criteria. The first criterion is to pick the window size so as to limit the number of timepoints required to represent each node waveform in a window. This puts a strict upper bound on the amount of storage needed for the waveforms, and thus allows the RELAX2 program to avoid dynamically managing waveform storage space. The second criterion is to try to pick the window size so that the convergence of the WR is rapid; in particular, so that the waveforms approach the correct solution in a uniform manner over the entire window. The RELAX2 program presently uses Algorithm 7.4 to determine the window size. At present, one-twentieth of the simulation interval is being used as an initial guess for the window size. Adding a simple critical-path analyzer to RELAX2 is being considering to provide a better initial guess.

SECTION 7.5 - PARTIAL WAVEFORM CONVERGENCE

If the WR algorithm is used to compute the time-domain behavior for very large circuits, often some pieces of the circuit will converge much more rapidly than others. The overall efficiency of the WR method can be improved if the waveforms that have already converged are not recomputed on every subsequent iteration.

To take advantage of partial waveform convergence requires a simple modification to Algorithm 4.1. Before giving the exact algorithm we present the following useful definition.

Algorithm 7.4 (RELAX2 Windowing Algorithm)
  starttime = Beginning of the window
  stoptime = End of the window
  endtime = End of user-defined simulation interval
  usedpts = Max. # of points used in the last window
  maxpts = Max. # of points in a waveform buffer
  prevwindow = Size of the window used in the previous iteration
  if (Not entirely converged in this window) then {
      if (usedpts ≥ maxpts) then {
          Shorten window if the waveforms overran storage buffers.
          stoptime = starttime + (prevwindow * maxpts * 0.7)/usedpts;
      }
      else if ((numiters mod 5) == 0) then {
              Half window size every five WR iterations.
          stoptime = prevwindow/2 + starttime;
      }
      else { *Else just do the same window again.*
          stoptime = starttime + prevwindow;
      }
  }
  else {
      starttime = stoptime;
      stoptime = starttime + (prevwindow * maxpts * 0.7)/usedpts;
  }
  ■

Definition 7.1: Let

$$\sum_{j=1}^{n} C_{ij}(v(t),\ u(t))\ v_j\ (t)\ =\ f_i(v(t),\ u(t))\quad v^i(0)\ =\ v_{i0} \qquad [7.9]$$

be the ith equation of the system in Eqn. (2.10). We say $v_j(t)$ is an input to this equation if there exist some $\alpha, t \in \mathbb{R}$ and $z, y \in \mathbb{R}^n$ such that $C_{ij}(z, u(t))y_j \neq C_{ij}(z + \alpha e_j, u(t))y_j$ or $f_i(z, u(t)) \neq f_i(z + \alpha e_j, u(t))$, where e_j is the jth unit vector. The input set for the ith equation is the set of $j \in [1, ..., n]$ such that $v_j(t)$ is an input ■.

The WR algorithm is then modified slightly using the notion of the set of inputs to a given differential equation.

Algorithm 7.5 - WR Algorithm with Partial Waveform Convergence

The superscript k denotes the iteration count, the subscript i denotes the component index of a vector and ε is a small positive number.

$k \leftarrow 0$

guess waveforms $v^0(t)$; $t \in [0,T]$ such that $v^0(0) = v_0$
       (for example, set $v^0(t) = v_0$, $t \in [0,T]$)

repeat {
    $k \leftarrow k + 1$
    foreach (i in N) {
        Partialflag $=$ *TRUE*
        if ($k = 1$) *Partialflag* $=$ *FALSE*
        For each ($j < i, j \in$ input set of v_i)
            if ($\max_{[0,T]} | v_j^k - v_j^{k-1} | > \varepsilon$) *Partialflag* $=$ *FALSE*
        For each ($j \geq i, j \in$ input set of v_i)
            if ($\max_{[0,T]} | v_j^{k-1} - v_j^{k-2} | > \varepsilon$) *Partialflag* $=$ *FALSE*
        if (*Partialflag* $=$ *TRUE*) $v_i^{k+1} = v_i^k$
        else solve

$$\sum_{j=1}^{i} C_{ij}(v_1^k, ..., v_i^k, v_{i+1}^{k-1}, ..., v_n^{k-1}, u)\dot{v}_j^k +$$

$$\sum_{j=i+1}^{n} C_{ij}(v_1^k, ..., v_i^k, v_{i+1}^{k-1}, ..., v_n^{k-1}, u)\dot{v}_j^{k-1} +$$

$$f_i(v_1^k, ..., v_i^k, v_{i+1}^{k-1}, ..., v_n^{k-1}, u) = 0$$

        for ($v_i^k(t)$; $t \in [0,T]$), with the initial condition $v_i^k(0) = v_{i_0}$

    }
} **until** ($\max_{1 \leq i \leq n} \max_{t \in [0,T]} | v_i^k(t) - v_i^{k-1}(t) | \leq \varepsilon$)
    that is, until the iteration converges.

■

SECTION 7.6 - EXPERIMENTAL RESULTS

The degree to which the WR algorithm improves circuit simulation efficiency can be traced to two properties of a circuit. The first, mentioned before, is the differences in the rates of change of voltages in the system, as this will determine how much efficiency is gained by solving the subsystems with independent integration timesteps. The second is the amount of coupling between the subsystems. If the subsystems are tightly coupled, then many relaxation iterations will be required to achieve convergence; and the advantage gained by solving each subsystem with its own timestep will be lost. To show this interaction for a practical example, we will use the RELAX2[13] program to compare the computation time required to simulate a 141-node CMOS memory circuit using standard direct methods and using the WR algorithm. In order to demonstrate the effect of tighter coupling, the CMOS memory circuit will be simulated using several values of a parameter XQC, which is the percent of the gate-oxide capacitance that is considered as gate-drain or gate-source overlap capacitance.

TABLE 7.3 - DIRECT VS WR FOR DIFFERENT COUPLINGS				
METHOD	XQC	TIMEPOINTS	# WR ITERS	CPU TIME
Direct	0.01	124,539	1	933s
WR	0.01	17,728	2.5	304s
Direct	0.05	122,988	1	945s
WR	0.05	19,199	3	410s
Direct	0.2	118,335	1	917s
WR	0.2	19,193	4	530s
Direct	0.33	115,233	1	895s
WR	0.33	19,315	6.5	707s

The results in Table 7.3 are exactly as expected. As the coupling increases, the number of WR iterations required increases, and the difference in the simulation times for WR and the direct method decreases.

The above example can be used to justify our claims about the nature of the efficiencies of the WR algorithm. We consider that a reasonable measure of the amount of work performed in a simulation is the sum, over all nodes in a given circuit, of the number of timepoints computed for each node. We refer to this as the sum of node-timepoints. Consider the number of node-timepoints computed by the direct method versus the number of node-timepoints computed by the WR method in the final iteration. By comparing these two numbers, a bound can be put on the maximum speed increase that can achieved by solving different subsystems using different timesteps (Note that we are only considering the number of timepoints computed by the WR method in the final iteration, because we are only interested in the number of timepoints needed to accurately represent the given waveform).

The total number of node-timepoints computed for each of the simulation cases of the memory circuit example is also given in Table 7.3. If most of the efficiency of a decomposition method stems from solving each of the subsystems with its own timestep, then the maximum improvement that could be gained from a decomposition integration method would be the ratio of the number of node-timepoints computed using the direct method compared to the number of timepoints computed in the final WR iteration. As can be seen from the Table 7.3, for the CMOS memory example this is approximately 6:1. In order to compute the actual efficiency of the WR method, the average number of WR iterations performed must be considered, because for each WR iteration the set of node-timepoints is recomputed. Then, if our above claims are correct, when the ratio of the number of node-timepoints for the direct method to the number of node-timepoints for WR is divided into the average number of relaxation iterations, the result should be almost equal to the ratio of WR computation time to direct method computation time. And, as Table 7.3 shows, it is.

In the above analysis we have ignored an important advantage of relaxation methods: they avoid large matrix solutions. This is a reasonable assumption for the above example because the matrix operations account for only a small percentage of the computations, even when direct methods are used. However, for much larger problems, of the order of several thousand nodes, the time to perform the large matrix solutions required by direct methods will dominate. In those cases WR methods should compare even more favorably because they avoid these large matrix solutions.

Finally, in Tables 7.4a and 7.4b, we compare the CPU time required to simulate several circuits using SPICE, RELAX2 with direct methods, and RELAX2 using the WR algorithm.

TABLE 7.4a – SPICE2G vs RELAX2 DIRECT vs WR* (Shichman-Hodges Mosfet model)				
Circuit	Devices	SPICE	DIRECT	WR
uP Control	232	1400s	90s	45s
Cmos Memory	621	10400s	995s	308s
4-bit Counter	259	4300s	540s	299s
Digital Filter	1082	18000s	1800s	520s
Encode-Decode	3295	115000s	5000s	1500s

*On Vax11/780 running Unix

TABLE 7.4b – TISPICE** vs RELAX2 DIRECT vs WR* (Yang-Chatterjee Mosfet model)				
Circuit	Devices	TISPICE	DIRECT	WR
Inverter Chain	250	439s	98s	38s
Eprom	348	694s	531s	349s
VHSIC Memory	625	15670s	15830s	12505s

*On Vax11/780 running VMS
**Texas Instrument's Circuit Simulation Program

CHAPTER 8 - PARALLEL WR ALGORITHMS

Exploiting parallel computation for circuit simulation is extremely important because the size of the circuits to which circuit simulation has been applied has grown at rate that far exceeds the increase in computational power due to technological improvement. The *only* way to keep pace with the increasing demand is to be able to apply many processors to the problem, and the number of processors that can be used must scale up with the size of the problem.

A variety of techniques for the parallel solution of ordinary differential equations have been examined in the literature[63]. For circuit simulation, four techniques have been applied. The SPICE2 program was rewritten to take advantage of the Cray computer's vector capability[48]; a parallel version of a similar direct method has been implemented on the Cosmic-Cube[69], a message-passing parallel computer; the Gauss-Jacobi form of the algebraic relaxation-Newton algorithm presented in Section 3.2 has been implemented on both a shared-memory computer, the Sequent Balance 3000[64], and ITM's Connection Machine[65]; and a version of the iterated timing analysis algorithm (Section 3.2) has been implemented on the BBN Butterfly[34].

In this chapter, the implementation of two WR-based parallel circuit simulation algorithms on a shared-memory computer will be described. We will start by presenting a brief overview of the aspects of a shared-memory computer that affect the algorithm implementation, and then we will describe the two parallel WR algorithms, one based on using a mixture of Gauss-Seidel and Gauss-Jacobi relaxation and the other based on pipelining the waveform computation. For each algorithm, experimental results will be presented.

SECTION 8.1 - AN OVERVIEW OF THE SHARED-MEMORY COMPUTER

When attempting to write efficient programs for serial computers knowledge about the specific details of the architecture is useful, but not essential. This is not the case for programming on a parallel computer. Specific details about the architecture can influence decisions about the implementation of an algorithm, and can even affect the choice of algorithm. Since the algorithms that will be described below were implemented on the Sequent Balance 8000, a shared-memory parallel computer, in this section we will describe those aspects of the architecture that affected the implementation of parallel versions of the WR algorithm. For a more detailed treatment of this subject, see[56].

The key problem in designing a parallel processor is that of communication between the processors. One simple approach is to design a parallel computer by gathering together many standard serial computers, and connecting them together with a comunication network. Usually such computers are referred to as *message-passing* parallel computers, because data are tranferred between the many processors by passing messages on the communication network. The disadvantage of such a system is that in order to move data from the memory of one processor into the memory of the second processor, both the transmitting and receiving processors must be involved.

Another approach to the problem of communicating between parallel processors is to redesign the memory system, so that the aggregate memory of all the processors is directly addressable by any one of the individual processors. Such a system is referred to as a *shared-memory* system because the processors are all sharing the single resource: the memory. The main advantages of a shared-memory machine is that it is not necessary to explicitly transfer data from one processor to another. When one processor needs data from another processor, it simply reads from the memory locations in which the other processor has written. This also allows for more dynamic algorithm structures, because it is not necessary to determine beforehand which processors will need the results of a given calculation. The disadvantages of the shared-memory computer are that all processors must contend for a single resource, the memory, and guaranteed synchronization between processors is not simple without special-purpose hardware.

One of the most important aspects of a shared-memory parallel computer is how the memory is distributed among the individual processors. There are two common choices: either each processor has a portion of the shared memory which it can access rapidly, and that others can access but not as quickly, or all the memory is centralized, and the many processors contend on an equal footing for access to it.

If the memory is distributed among the processors, then a parallel algorithm will perform better if the data for the computation can be partitioned so that each processor performs computations using only the data in its own portion of the shared memory. It is usually the case that by partitioning the data so that each of the processors can only work on an exclusive portion of a large problem, some of the parallelism of a given algorithm will no longer be exploitable and parallel efficiency will be lost. For example, if at a certain point in the process of solving a large problem, several calculations that could be performed concurrently all require data from the same partition, those calculations will be performed serially. If, at the same time, there are no calculations to be performed using data from another partition, a processor will be idle.

A way of eliminating the loss of parallelism at the cost of complicating the control structure of the program is to have each processor use a priority scheme. In such a scheme, each processor attempts to perform calculations using data in its own partition, and then if there are none to be performed, the processor will atttempt to perform calculations using data from other partitions.

Clearly, when using a shared-memory computer with distributed memory, the trade-offs of faster memory access, loss of parallelism, and more complicated control structure must be examined carefully(For an example in circuit simulation see[34]).

The memory on the parallel computer used for these parallel WR experiments is centralized, with all the processors contending for one large shared memory. For such an architecture, there is no advantage to partitioning the data for a large problem among the processors, as they will still have to contend for the same centralized memory pool. For this reason, the algorithms presented below ignore the issue of partitioning the data among many processors.

In order to avoid the obvious bottleneck created by having many processors contend for data out of the same central memory, most implementations of shared-memory computers that use centralized memory attempt to reduce this conflict by including a large cache memory within each of the processors. As with any cache memory scheme, these caches attempt to exploit locality of reference[70]; that is, each of the processors are actively using only a small number of data. Since these data will probably be available from the cache, most memory accesses will not generate a request to the main central memory.

Using caches on a parallel computer is not as straightforward as on a serial computer. Since there are many caches, and they are all supposed to contain a copy of the data in the central memory, and any processor can write in any memory location, it is possible for the caches to lose consistency. That is, the contents of a cache may not reflect the current contents of the central memory. For example, if the contents of memory location A is in both the cache for processor 1 and the cache for processor 2, and processor 1 updates A, then the data in the cache for processor 2 will be incorrect.

As the example demonstrates, even if the central memory is updated whenever a processor updates a location contained in its cache, a cache inconsistency can occur within the cache of another processor. There are a variety of schemes for avoiding this problem[62], but we will only mention the technique applied in the computer used for experimentation. The scheme is simple: all the caches monitor all the writes to central memory from any of the processors. If a cache contains a location being written to by any of the other processors, it updates its own copy of the data in the given location. By snooping on the writes to central memory, each cache assures that it has the most current data.

The snooping-cache consistency strategy has a particularly useful implication. It is frequently necessary to have one processor wait for another processor to finish a computation. If the computing processor is to change the contents of a location in memory when finished, the second processor can continuously test that location to determine when the computing processor is finished. Normally, this is a poor approach for a parallel environment, because the waiting processor will be continuously reading from the central memory and generating excess memory traffic. If many processors are waiting for the completion of one processor's computation, this excess traffic can become enormous and slow the computing processor which has to contend with the excess traffic. If the cache architecture described above is used, the excess traffic is eliminated. Each waiting processor will keep rereading a location which will be in its own cache, and will therefore not be generating *any* central memory traffic. When the computing processor finishes, each of the other processor caches will spot the write to the monitored location in central memory and each cache will update its own copy of the data. The waiting processors will therefore be made immediately aware of the completion of the computing processor, but will not have impeded the progress of the computing processor by generating excess memory traffic.

The last aspect of the parallel computer architecture that we will consider is that of mutual exclusion or *locking*. In almost all parallel programs there are critical sections that must be performed serially; that is, only one processor should be executing the section at a time. The usual mechanism for insuring this is the *test-and-set* instruction. If a processor executes a *test-and-set* instruction on a given location in memory, the contents of the location are returned to the processor and simultaneously, if the location was not set, it is set.

The mechanism can be used to perform locking as follows. A particular location in memory is used as the lock. If a processor is about to execute a critical section of a parallel program, it first executes a test-and-set on the lock location. If the result indicates that the location was not set, then the processor can safely execute the critical section, and clear the lock location when finished. If the result indicates that the lock was already set, the processor must wait until the lock becomes clear and then retry the test-and-set.

SECTION 8.2 - MIXED SEIDEL/JACOBI PARALLEL WR ALGORITHM

An obvious way of parallelizing WR is to use the Gauss-Jacobi version of WR. In this algorithm, the relaxation makes use of the waveforms computed at the previous iteration for all the subsystems. In this case, all the subsystems can be analyzed independently by different processors. One of the difficulties in applying this algorithm is that MOS digital circuits are highly directional and, as mentioned in Section 7.2, if this directionality is not exploited slow convergence may result. For example, consider applying WR to compute the transient response of a chain of inverters. If the first inverter's output is computed first, and the result is used to compute the second inverter's output, which is then used for the third inverter, etc., the resulting waveforms for this first iteration of the WR algorithm will be very close to the correct solution. However, if the second and third inverter outputs are computed in parallel with the first inverter's output, the results will not be close to the correct solution because no reasonable guess for the second and third inverter inputs will be available. For this reason, after partitioning, the RELAX2.3 program orders the subcircuits so that the directionality of the circuit is followed as closely as possible.

Following a strict ordering of the relaxation computation (Gauss-Seidel) does not allow for computing entire waveforms in parallel, and computing the next iteration waveforms for every subcircuit at once (Gauss-Jacobi) allows for substantial parallelism, but is not very efficient (converges more slowly). In order to preserve the efficiency of the Gauss-Seidel algorithm and allow for some of the parallelism of Gauss-Jacobi, a mixed approach can be employed. The mixed approach is based on the observation that large digital circuits contain many subsystems that can be computed in parallel without slowing the convergence. This is because large digital circuits tend to be wide. Rather than being like a long chain of gates, they are like many parallel chains, with some interaction between the chains. For this reason, it is possible to order the computation so that subcircuits in parallel "chains" can be computed in parallel, but the serial dependence inside a chain is preserved. This will not allow for as much parallelism as the Gauss-Jacobi scheme, but should preserve most of the efficiency of the Gauss-Seidel scheme.

In Algorithm 8.1, we present a probabilistic approach to attempting to follow the ordering of the subcircuits. The algorithm is set up by establishing both the space in shared memory for storage of the iteration waveforms, and a buffer or queue with the list of subcircuits in the order derived from Algorithm 7.3. Each of the processors then begins by taking a subcircuit from the queue and computing the subcircuit's output waveforms using the newest available external waveforms. When the waveform computation is completed, the subcircuit is temporarily discarded and the processor takes the next subcircuit off the queue. This continues until the queue is exhausted and all the processors are finished. Then the queue is reset, and the processors all start picking up subcircuits again.

This algorithm is probabilistic in the sense that there is no guarantee that the transient computation for a given subcircuit will be finished before its output is needed by another subcircuit which is strongly serially dependent on the first subcircuit's output. It is likely that the given subcircuit's output will have been computed if the circuit is very wide (there are a large number of parallel chains) compared to the number of processors. In addition, since all the subcircuit outputs must be computed before any subcircuit's output is recomputed, no subcircuit will be more than one iteration behind.

Algorithm 8.1 - (Jacobi/Seidel-based Parallel WR)

    Initialization. Both subcircuits and waveforms in shared-memory.
    queue = ordered__list__of__subcircuits
    while (all__converged == FALSE) { *Parallel iteration loop. All processors execute.*
        if (processor == 1) {
            reset__queue()
            idle__count = 0
        }
        while (idle__count ≠ number of processors) {
            while (test-and-set(queuelock) == set) { *Tight loop waiting for queue to unlock.* }
            Queue is locked, get next subcircuit
            NextSub = Get__next__queue__entry()
            if (NextSub == NULL) {
                increment(idle__count)
                clear(queuelock)
            }
            else { *There is another subcircuit on the queue.*
                clear(queuelock)
                Compute__Subcircuit__Waveforms(NextSub)
                Check__Waveform__Convergence(NextSub)
            }
        }
    }
■

Note that the attributes of the parallel architecture have been considered in Algorithm 8.1. Since the machine is a centralized shared-memory machine, the data describing the subcircuits and the computed waveforms are left in shared memory, to be accessed as needed. Also note that each of the processors waits for the queue to be free by examining the lock variable in a tight loop. As mentioned above, this exploits the nature of the cache consistency strategy. Finally, in this case it is not necessary to separately control access to the waveforms. Since the waveforms will only be written as a result of the computations performed on their associated subcircuits, and a waveform is associated with only one subcircuit, the mutual exclusion of the subcircuit queue will prevent waveform writes from colliding.

SECTION 8.3 – TIMEPOINT-PIPELINING WR ALGORITHM

It is possible to parallelize the WR algorithm while still preserving a strict ordering of the computation of the subcircuit waveforms (Gauss-Seidel) by pipelining the waveform computation. In this approach, one processor starts computing the transient response for a subcircuit. Once a first timepoint is generated, a second processor begins computing the first timepoint for the second subcircuit, while the first processor computes the second timepoint for the first subcircuit. On the next step a third processor is introduced to compute the first timepoint for the third subcircuit, and so on.

Conceptually, the operations of a given processor in a parallel timepoint-pipelining algorithm are quite simple. The algorithm is set up by establishing both the space in shared memory for storage of the iteration waveforms, and a buffer or queue with the list of subcircuits. Each of the processors then starts by taking a subcircuit from the queue. The individual processors examine their respective subcircuit's external waveforms to see if the waveform values needed to compute the next integration timestep are available. If so, the next timestep for the subcircuit is computed. Otherwise, the subcircuit is returned to the queue and the processor tries again with another subcircuit from the queue. As timepoints are computed, more of the subcircuits will have the information needed to compute their own timepoints.

As one might expect, a practical timepoint-pipelining algorithm is more complicated that the conceptual algorithm. Perhaps the most obvious difficulty is that there is a tremendous overhead in having every processor search through all the subcircuits to find one of the few for which a timepoint can be computed.

It is possible to reduce the number of candidate subcircuits a processor must search by only considering those subcircuits for which at least one of the external waveforms has more timepoints than it had when the subcircuit was last processed. Clearly, this will avoid having the processors continuously rechecking subcircuits for which no new information is available, and therefore no new timestep could be computed.

This kind of selective-search algorithm can be implemented by altering the way in which the queue of subcircuits is used. When a processor discerns that it is not possible to compute a new timepoint for a subcircuit, instead of returning the subcircuit to the queue, the subcircuit is temporarily discarded. If a processor succeeds in computing a timepoint for a subcircuit, those subcircuits that are connected to the given subcircuit, referred to as the *fanouts* of the subcircuit, are added to the queue (Of course, any of the fanouts that are already on the queue are not duplicated). In this way, the only subcircuits that will be on the queue are those for which it is likely that the waveform values needed to compute a next timepoint will be available.

Another aspect of the timepoint-pipelining algorithm that increases the exploitable parallelism at the cost of slightly complicating the algorithm is to allow the timepoint pipelining to extend across iteration boundaries. For example, consider a chain of two inverters, and assume that it takes two timesteps to compute each of the inverter outputs. As before, the second timestep of the first inverter can be computed in parallel with the first timestep of the second inverter. Then, while the second timestep of the the second inverter is being computed, there is enough information to compute the first timestep of the first inverter for the *second* WR iteration.

This enhancement does not really complicate the conceptual algorithm until one considers when to stop. For a long chain of inverters, allowing the pipelining to extend across iteration boundaries can easily allow the first inverter to be many iterations ahead of the last inverter. Since WR convergence can only be determined when all the waveforms for a given iteration have been computed, it may well be that the WR iteration being computed for the first inverter is many iterations beyond what is necessary to achieve satisfactory convergence. The difficulty is that this fact will not be discovered until much later, when all inverter outputs have been computed in the iteration for which satisfactory convergence was achieved.

This is not a disastrous problem; the algorithm will still produce correct solutions, but unnecessary computations will be performed and efficiency will be lost. The unnecessary computations are simple to avoid. Subcircuits can be postponed from starting on iteration $N+1$ until nonconvergence of some waveform of iteration N has been detected. It is, of course, important to discover as quickly as possible if it will be necessary to compute iteration $N+1$, so that the pipelining of that iteration can begin. For this reason, in the timepoint-pipelining algorithm presented below, convergence is checked on a timepoint-by-timepoint basis, that is, immediately after each timepoint is computed.

SECTION 8.4 - PARALLEL ALGORITHM TEST RESULTS

As mentioned above, the two algorithms were implemented on a nine-processor configuration of the Sequent Balance 8000 computer (larger configurations are available). The results from several experiments for the two algorithms are given in Tables 8.1 and 8.2. As the results from the EPROM and micro-processor control circuit indicate, the timepoint-pipelining algorithm makes much more efficient use of the available processors. In fact, as Table 8.2 shows, the timepoint-pipelining algorithm running on the Balance 8000 runs substantially faster than the serial WR algorithm running on a Vax/780.

A second point should be made about the timepoint-pipelining examples. It can be seen that the speed-up is not linear to nine processors, but starts to drop off. This seems surprising given the size of the examples, but not when the type of circuit being simulated is considered. For the biggest example, the CMOS

Algorithm 8.2 - (Timepoint-Pipelining WR Algorithm)
Initialization. Both subcircuits and waveforms in shared memory.
queue = ordered__list__of__subcircuits
idle__count = 0
Max__iter__so__far is one greater than the last iteration for which nonconvergence was detected.
max__iter__so__far = 1
Parallel iteration loop. All processors execute.
while (idle__count ≠ # of processors) { ≥ 1 *processor still computing.*
  while (test-and-set(queuelock) == set) { *Wait for queue to unlock.* }
  Queue is locked, get next subcircuit in the queue for which the work that might be performed on it is for an iteration that is no more than one beyond the maximum iteration for which nonconvergence has been detected.
  NextSub = Get__next__queue__entry(max__iter__so__far)
  if (NextSub == NULL) {
    increment(idle__count)
    clear(queuelock)
  }
  else {
    There is a subckt on the queue whose iteration is not beyond max__iter__so__far.
    clear(queuelock)
    repeat { *Compute all timepoints possible with available waveform values.*
      Check that needed external values are available
      cando = Check__for__next__step(NextSub)
      if (cando == TRUE) {
        Compute__Next__Step(NextSub)
        converged = Check__Step__Convergence(NextSub)
        if ((converged == FALSE) and
          (NextSub.iter__count == max__iter__so__far)) {
        Keep max__iter__so__far ahead of the nonconverged iterations.
        increment(max__iter__so__far)
        }
        enqueue__fanouts(NextSub)
      }
    } until (cando == FALSE)
  }
}

■

RAM, the partitioning algorithm produces approximately 75 subcircuits, and this would indicate that a speed-up of 75 should be obtainable, or at least approachable. This ignores one of the features of the WR algorithm: only active portions of the circuit are participating in the computation. For digital circuits, this is usually less than ten percent of the circuit. This implies that for the CMOS RAM example, over any given interval, roughly seven subcircuits are active and involved in the computation, and therefore only a speed-up of seven could be expected.

Table 8.1 - G-S/G-J WR VS # OF PROCESSORS					
Circuit	FET's	1	3	6	9
uP Control	66	595	338	270	259
Eprom	348	512	317	286	266

Table 8.2 - TIMEPOINT PIPELINING WR VS # OF PROCESSORS					
Circuit	FET's	1	3	6	9	VAX/780
uP Control	116	704	247	159	149	240
Eprom	348	745	265	185	182	212
Cmos Ram	428	3379	1217	642	496	960

REFERENCES

[1] C. William Gear, *Numerical Initial Value Problems for Ordinary Differential Equations*, Prentice-Hall Inc., Englewood Cliffs, New Jersey, 1974.

[2] L.W. Nagel, "SPICE2: A computer program to simulate semiconductor circuits," Electronics Research Laboratory *Report No. ERL-M520*, University of California, Berkeley, May 1975.

[3] W. T. Weeks, A. J. Jimenez, G. W. Mahoney, D. Mehta, H. Qassemzadeh, and T. R. Scott, "Algorithms for ASTAP -- A Network Analysis Program," *IEEE Trans. on Circuit Theory*, Vol. CT-20, No. 6, November 1973, pp. 628-634.

[4] K. Sakallah and S.W. Director, "SAMSON2: An Event Driven VLSI Circuit Simulator," *IEEE Trans. on Computer-Aided Design*, Vol. CAD-4, No. 4, October 1985, pp. 668-684.

[5] G. De Micheli, A. Sangiovanni-Vincentelli, and A.R. Newton, "New Algorithms for the Timing Analysis of Large Circuits," *Proc. Int. Symp. on Circuits and Systems*, Houston, Texas, 1980.

[6] G. De Micheli and A. Sangiovanni-Vincentelli, "Characterization of Integration Algorithms for the Timing Analysis of MOS VLSI Circuits," *Int. J. of Circuit Theory and Applications*, Vol. 10, October 1982, pp. 299-309.

[7] B.R. Chawla, H.K. Gummel, and P. Kozah, "MOTIS - an MOS Timing Simulator," *IEEE Trans. on Circuits and Systems*, Vol. CAS-22, December 1975, pp. 901-909.

[8] C. F. Chen and P. Subramanyam, "The Second Generation MOTIS Timing Simulator -- An Efficient and Accurate Approach for General MOS Circuits" *Proc. Int. Symp. on Circuits and Systems*, Montreal, Canada, May 1984.

[9] Keepin, W. N., "Multirate Integration of Two Time-Scale Systems," *Ph.D. Dissertation*, University of Arizona, 1980.

[10] C. William Gear, "Automatic Multirate Methods for Ordinary Differential Equations," *Information Processing 80*, North-Holland Pub. Co., Amsterdam, 1980.

[11] E. Lelarasmee, A. E. Ruehli, and A. L. Sangiovanni-Vincentelli, "The Waveform Relaxation Method for Time Domain Analysis of Large Scale Integrated Circuits," *IEEE Trans. on CAD of IC and Systems*, Vol. 1, No. 3, July 1982, pp. 131-145.

[12] E. Lelarasmee, "The Waveform Relaxation Method for the Time Domain Analysis of Large Scale Nonlinear Dynamical Systems," *Ph.D. dissertation*, University of California, Berkeley, 1982.

[13] E. Lelarasmee and A. Sangiovanni-Vincentelli, "Some New Results on Waveform Relaxation Algorithms for the Simulation of Integrated Circuits," *Proc. Int. Symp. on Large Scale Systems*, Virginia Beach, Virginia, October 1982.

[14] E. Lelarasmee and A. Sangiovanni-Vincentelli, "Relax: A New Circuit Simulator for Large Scale MOS Integrated Circuits," *Proc. 19th Design Automation Conference*, Las Vegas, Nevada, June 1982.

[15] P. Defebve, J. Beetem, W. Donath, H.Y. Hsieh, F. Odeh, A.E. Ruehli, P.K. Wolff, Sr., and J. White, "A Large-Scale Mosfet Circuit Analyzer Based on Waveform Relaxation," *Proc. Int. Conf. on Computer Design*, Rye, New York, October 1984.

[16] C. H. Carlin and A. Vachoux, "On Partitioning for Waveform Relaxation Time-Domain Analysis of VLSI Circuits," *Proc. Int. Symp. on Circuits and Systems*, Montreal, Canada, May 1984.

[17] J. White and A. Sangiovanni-Vincentelli, "Relax2: A Modified Waveform Relaxation Approach to the Simulation of MOS Digital Circuits," *Proc. Int. Symp. on Circuits and Systems*, Newport Beach, California, May 1983.

[18] J. White and A. Sangiovanni-Vincentelli, "Relax2.1 - A Waveform Relaxation Based Circuit Simulation Program," *Proc. Int. Custom Integrated Circuits Conference*, Rochester, New York, June 1984.

[19] M. Guarini and O. A. Palusinski, "Integration of Partitioned Dynamical Systems using Waveform Relaxation and Modified Functional Linearization," *Summer Computer Simulation Proceedings,* Vancouver, Canada, July 1983.

[20] W.M.G. van Bokhoven, "An Activity Controlled Modified Waveform Relaxation Method," *Proc. Int. Symp. on Circuits and Systems,* Newport Beach, California, May 1983.

[21] J. M. Ortega and W.C Rheinbolt, *Iterative Solution of Nonlinear Equations in Several Variables,* Academic Press, New York, 1970.

[22] O. Palusinski, "Continuous Expansion Methods in Computer Aided Circuit Analysis," *Proc. Int. Custom Integrated Circuits Conference,* Rochester, New York, June 1984.

[23] J. Kaye and A. Sangiovanni-Vincentelli, "Solution of piecewise linear ordinary differential equations using waveform relaxation and Laplace transforms," *IEEE Trans. on Circuits and Systems,* Vol. CAS-31, June 1983, pp. 353-357.

[24] W.K. Chia, T.N. Trick, and I.N. Haij, "Stability and Convergence Properties of Relaxation Methods for Hierarchical Simulation of VLSI Circuits," *Proc. Int. Symp. on Circuits and Systems,* Montreal, Canada, May 1984.

[25] F. Odeh and D Zein, "A Semidirect Method for Modular Circuits," *Proc. Int. Symp. on Circuits and Systems,* Newport Beach, California, May 1983.

[26] G. Guardabassi, "Subsystemwise Simulation of Discrete-Time Interconnected Dynamical Systems," Electronics Research Laboratory *Report No. ERL-M82/8,* University of California, Berkeley, February 1982.

[27] H. A. Antosiewicz, "Newton's method and boundary value problems," *J. of Computer System Science,* Vol. 2, No. 2, 1968, pp. 177-202.

[28] R. S. Varga, *Matrix Iterative Analysis,* Prentice-Hall Inc., Englewood Cliffs, New Jersey, 1962.

[29] F. Odeh, A. Ruehli, and C. H. Carlin "Robustness Aspects of an Adaptive Waveform Relaxation Scheme," *Proc. Int. Conf. on Computer Design,* Port Chester, New York, October 1983.

[30] J. More, "Nonlinear Generalization of Matrix Diagonal Dominance with Applications to Gauss-Seidel Iterations" *SIAM J. Numerical Analysis,* Vol. 9, No. 2, June 1972.

[31] J. White and A.L. Sangiovanni-Vincentelli, "Partitioning Algorithms and Parallel Implementations of Waveform Relaxation Algorithms for Circuit Simulation," *Proc. Int. Symp. on Circuits and Systems,* Kyoto, Japan, June 1985.

[32] U. Miekkala and O. Nevanlinna, "Convergence of Dynamic Iteration Methods for Initial Value Problems," *Report MAT-A230,* Institute of Mathematics, Helsinki University of Technology, Finland, 1985.

[33] R. A. Saleh, J. E. Kleckner and A. R. Newton, "Iterated Timing Analysis and SPLICE1," *Proc. Int. Conf. on Computer-Aided Design,* Santa Clara, California, September 1983.

[34] J. T. Deutsch, "Algorithms and Architecture for Multiprocessor-Based Circuit Simulation," *Ph.D. Dissertation,* University of California, Berkeley, 1985.

[35] W. Rudin, *Functional Analysis,* McGraw Hill, New York, 1969.

[36] C. A. Desoer and E. S. Kuh, *Basic Circuit Theory,* McGraw Hill, New York, 1969.

[37] R. Courant and D. Hilbert, *Partial Differential Equations,* Vol. 2 of *Methods of Mathematical Physics,* Interscience Pub. Inc., New York, 1962.

[38] C. W. Ho, A. E. Ruehli, and P. A. Brennan, "The Modified Nodal Approach to Network Analysis," *IEEE Trans. on Circuits and Systems,* Vol. CAS-22, June 1975, pp. 504-509.

[39] J. K. Hale, *Ordinary Differential Equations,* John Wiley and Sons Inc., New York, 1969.

[40] L. Chua and P. Lin, *Computer-Aided Analysis of Electronic Circuits: Algorithms and Computational Techniques,* Prentice-Hall Inc., Englewood Cliffs, New Jersey, 1975.

[41] P. Yang, B.D. Epler, and P. K. Chatterjee, "An Investigation of the Charge Conservation Problem for MOSFET Circuit Simulation," *IEEE J. of Solid-State Circuits,* Vol. SC-18, No. 1, Feb. 1983, pp. 128-138.

[42] G. Dahlquist, "A Special Stability Problem for Linear Multistep Methods," *BIT,* No. 3, 1963, pp. 27-43.

[43] E. M. Purcell, *Electricity and Magnetism.* McGraw-Hill, New York, 1965.

[44] K. S. Kundert, "Sparse Matrix Techniques," *Circuit Analysis, Simulation and Design. Vol. 1.,* edited by A. E. Ruehli, North-Holland, to be published in 1986.

[45] A. R. Newton, "The Simulation of Large Scale Integrated Circuits," Electronics Research Laboratory, *Memo No. ERL-M78/52,* University of California, Berkeley, July 1978.

[46] R. K. Brayton and C. Conley, "Some Results on the Stability and Instability of the Backward Differentiation Methods with non-uniform Time steps," *Topics in Numerical Analysis,* Proc. Royal Irish Academy Conf., Academic Press, New York, 1972.

[47] A. L. Sangiovanni-Vincentelli, "On Decomposition of Large Scale Systems of Linear Algebraic Equations," *Proc. of JACC,* Denver, Colorado, June 1979.

[48] A. Vladimirescu and D. O. Pederson, "Performance Limits of the CLASSIE Circuit Simulation Program," *Proc. Int. Symp. on Circuits and Systems,* May 1982.

[49] J. A. George, "On Block Elimination for Sparse Linear Systems," *SIAM J. of Numerical Analysis,* Vol. 11, 1974, pp. 585-603.

[50] W. L. Engl, R. Laur, and H. K. Dirks, "MEDUSA -- A Simulator for Modular Circuits," *IEEE Trans. on CAD of IC and Systems,* Vol. CAD-1, No. 2, April 1982, pp. 85-93.

[51] P. Yang, I. N. Hajj, and T. N. Trick, "SLATE: A Circuit Simulation Program with Latency Exploitation and Node Tearing," *Proc. Int. Conf. on Circuits and Computers,* New York, October 1980.

[52] N. B. G. Rabbat, A. L. Sangiovanni-Vincentelli, and H. Y. Hsieh, "A Multi-level Newton Algorithm with Macromodelling and Latency for the Analysis of

Large-Scale Nonlinear Circuits in the Time Domain," *IEEE Trans. on Circuits and Systems,* Vol. CAS-26, September 1979, pp. 733-741.

[53] A.R. Newton and A. L. Sangiovanni-Vincentelli, "Relaxation-Based Circuit Simulation," *IEEE Trans. on ED,* Vol. ED-30, No. 9, Sept. 1983, pp. 1184-1207. also *SIAM J. of Scientific and Stat. Computing,* Vol. 4, No. 3, September 1983. also *IEEE Trans. on CAD of IC and Systems,* Vol. CAD-3, No. 4, October 1984, pp. 308-330.

[54] W. Kahan, Private Notes, 1975.

[55] T. Huang, "Generalization of the Implicit-Implicit-Explicit Method for Floating Capacitors," *Master's Thesis,* University of California, Berkeley, 1983.

[56] A. Gottlieb, R. Grishman, C. P. Kruskal, K. P. McAuliffe, L. Rudolph, and M. Snir, "The NYU Ultracomputer -- Designing an MIMD Shared Memory Parallel Computer," *IEEE Trans. on Computers* Vol. C-32, No. 2, Feb. 1983, pp. 175-188.

[57] J. Kleckner, "Advanced Mixed-Mode Simulation Techniques," *PhD. Thesis,* University of California, Berkeley, 1984.

[58] C. De Boor, *A Practical Guide to Splines,* Springer-Verlag, New York, 1978.

[59] O. Nevanlinna and F. Odeh, "Multiplier Techniques for Linear Multistep Methods," *Numer. Funct. Anal. and Optimiz.* Vol. 3, No. 4, 1981, pp. 377-423.

[60] J. White, N. Weiner, D. Webber, K. Kundert, P. Moore, R. Saleh, A. L. Sangiovanni-Vincentelli, and A. R. Newton, "RELAX2.3 User's Manual," to appear.

[61] D. Mitra, "Chaotic, Asynchronous Relaxations for the Numerical Solution of Differential Equations by Parallel Processors," *AT&T Bell Laboratories Memorandum,* October 1984.

[62] J. R. Goodman, "Using Cache Memory to Reduce Processor-Memory Traffic," *Proc. 10th Annual Symp. on Computer Architecture,* June 1983, pp. 124-131.

[63] W. Miranker, "A Survey of Parallelism in Numerical Analysis," *SIAM Review,* No. 13, 1971, pp 524-547.

[64] R. Saleh, Private Communication, 1985.

[65] D. Webber, Private Communication, 1985.

[66] J. White, R. Saleh, A. Sangiovanni-Vincentelli, and A. R. Newton, "Accelerating Relaxation Algorithms for Circuit Simulation Using Waveform Newton, Iterative Stepsize Refinement, and Parallel Techniques," *Proc. Int. Conf. on Computer-Aided Design,* Santa Clara, California, October 1985.

[67] J. Rubinstein, P. Penfield, Jr., and M. Horowitz, "Signal Delay in RC Tree Networks," *IEEE Trans. on CAD of IC and Systems,* Vol. CAD-2, No. 3, July 1983, pp. 202-211.

[68] G. D. Hachtel, R. K. Brayton, and F. G. Gustavson, "The Sparse Tableau Approach to Network Analysis and Design," *IEEE Trans. on Circuit Theory,* Vol. CT-18, January 1971, pp. 101-113.

[69] C. L. Seitz, "The Cosmic-Cube," *Comm. of the ACM,* January 1985, pp. 22-33.

[70] D. P. Siewiorek, C. G. Bell, and A. Newell, *Computer Structures: Principles and Examples,* McGraw Hill, New York, 1982.

[71] J. R. Duley and D. L. Dietmeyer, "A Digital Design System Language (DDL)," *IEEE Trans. on Computers,* Vol. C-17, No. 9, Sept. 1968, pp. 850-860.

[72] C. Y. Chu, "An ALGOL-like Computer Design Language," *Comm. of the ACM,* Vol. 8, No. 10, October 1965, pp. 607-615.

[73] A number of references may be found in *IEEE Trans. on Computers,* Vol. 7, No. 12, December 1974.

[74] N. Tanabe, H. Nakamura, and K. Kawakita, "MOSTAP: An MOS Circuit Simulator for LSI," *Proc. Int. Symp. on Circuits and Systems,* Houston, Texas, April 1980.

[75] G. Arnout, and H. De Man, "The Use of Threshold Functions and Boolean-Controlled Network Elements for Macromodelling of LSI Circuits," *IEEE J. of Solid-State Circuits,* Vol. SC-13, June 1978, pp. 326-332.

[76] A. R. Newton, "Techniques for the Simulation of Large-Scale Integrated Circuits," *IEEE Trans. on Circuits and Systems,* Vol. CAS-26, No. 9, Sept. 1979, pp. 741-749.

[77] K. Sakallah and S. Director, "An Event Driven Approach for Mixed Gate and Circuit-Level Simulation," *Proc. Int. Symp. on Circuits and Systems,* 1982, pp. 1194-1197.

[78] N. N. Nham, et. al., "The Mixed Mode Simulator," *Proc. 17th Design Automation Conference,* Minneapolis, Minnesota, June 1980.

[79] S. P. Fan, M. Y. Hsueh, A. R. Newton, and D. O. Pederson, "MOTIS-C A new circuit simulator for MOS LSI circuits," *Proc. Int. Symp. on Circuits and Systems,* April 1977.

[80] G. De Micheli, A. R. Newton, and A. Sangiovanni-Vincentelli, "Symmetric Displacement Methods for the Timing Analysis of Large Scale Circuits" *IEEE Trans. on CAD of IC and Systems,* Vol. CAD-2, No. 3, July 1983, pp. 167-180.

[81] J. White, A. S. Vincentelli, F. Odeh, and A. Ruehli, "Waveform Relaxation: Theory and Practice," *Trans. of The Society for Computer Simulation,* Vol. 2, No. 2, June 1985, pp. 95-134.

INDEX